Advances in
Energy Systems and Technology

Volume 5

Advances in
Energy Systems and Technology

Volume 5

Edited by
PETER L. AUER

Sibley School of Mechanical and
Aerospace Engineering
Cornell University
Ithaca, New York

DAVID DOUGLAS

Electric Power Research Institute
Palo Alto, California

1986

ACADEMIC PRESS, INC.
Harcourt Brace Jovanovich, Publishers

Orlando   San Diego   New York   Austin
London   Montreal   Sydney   Tokyo   Toronto

ACADEMIC PRESS, INC.
Orlando, Florida 32887

United Kingdom Edition published by
ACADEMIC PRESS INC. (LONDON) LTD.
24–28 Oval Road, London NW1 7DX

LIBRARY OF CONGRESS CATALOG CARD NUMBER: 78-4795

ISBN 0–12–014905–2

PRINTED IN THE UNITED STATES OF AMERICA

86 87 88 89   9 8 7 6 5 4 3 2 1

Contents

THE DEMAND FOR HOME INSULATION: A STUDY IN
THE HOUSEHOLD DEMAND FOR CONSERVATION
    J. Daniel Khazzoom

Contributors

Numbers in parentheses indicate the pages on which the authors' contributions begin.

Arnold P. Fickett (1), Advanced Conversion and Storage Department, Electric Power Research Institute, Palo Alto, California 94303

N. D. Kaushika (75), Centre for Energy Studies, Indian Institute of Technology, Hauz Khas, New Delhi 110 016, India

J. Daniel Khazzoom (167), Department of Economics, University of California, Berkeley, California 94720

Fuel Cells for Electric Utility Power Generation

Arnold P. Fickett

Advanced Conversion and Storage Department
Electric Power Research Institute
Palo Alto, California

1

I. INTRODUCTION

A fuel cell is an electrochemical device that continuously converts the chemical energy of a fuel (and oxidant) to electrical energy. The essential difference between a fuel cell and a storage battery is the continuous nature of the fuel and oxidant supply.

A simple fuel cell is illustrated in Fig. 1. Two catalyzed carbon rods (electrodes) are immersed in an electrolyte (acid in this illustration) and separated by a gas barrier. The fuel (in this case, hydrogen) is bubbled across the surface of one electrode while the oxidant (in this case, oxygen from ambient air) is bubbled across the other electrode. When

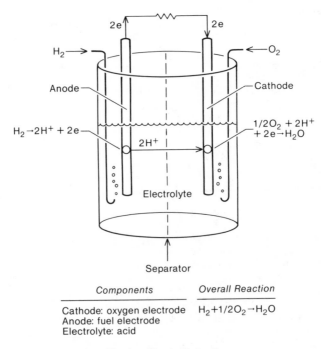

Fig. 1. Simple fuel cell.

the rods are electrically connected through an external load, the following events occur.

1. The hydrogen dissociates on the catalytic surface of the fuel electrode, forming hydrogen ions and electrons.
2. The hydrogen ions migrate through the electrolyte and separator (gas barrier) to the catalytic surface of the oxygen electrode.
3. Simultaneously, the electrons move through the external circuit to the oxygen electrode.
4. The oxygen, hydrogen ions, and electrons combine at the oxygen electrode to form water.

The net reaction is that of hydrogen and oxygen producing water and electrical energy; electrical energy is produced by the flow of electrons through the external circuit.

A practical fuel cell power plant, depicted in Fig. 2, will comprise at least three basic subsystems:

1. A power section, which consists of one or more fuel cell stacks: Each stack contains many individual fuel cells connected in series to produce a stack output ranging from a few to several hundred volts (dc).
2. A fuel subsystem that manages the fuel supply to the power section: This subsystem can range from simple flow controls to a complex fuel-processing facility.
3. A power conditioner that converts the output from the power section to the power type and quality required by the application: This subsystem could range from a simple voltage control to a

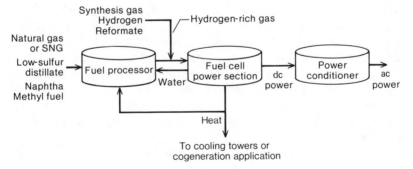

Fig. 2.   Fuel cell power plant.

sophisticated device that would convert the dc power to an ac power output.

In addition, a fuel cell power—depending on its size, type, and sophistication—may require controls of varying complexity and an oxidant subsystem, as well as thermal and fluid management subsystems.

Since they produce power by an electrochemical rather than a thermal cycle, fuel cells are not subject to the Carnot cycle limitation of thermal machines and thus offer the potential for highly efficient conversion of chemical to electrical energy. Furthermore, the efficiency is essentially independent of size; small power plants operate nearly as efficiently as large ones.

Fuel cell power plants are quiet and clean, the by-products being water, carbon dioxide, and nitrogen. Thus, they can be considered for applications where noxious emissions or noise would be objectionable or where water is unavailable.

Fuel cell power plants offer considerable flexibility. They can be configured to use a wide variety of fuels and produce a wide range of dc or ac power outputs.

These and other advantages have led to the development of fuel cells for numerous applications, including space, underwater, transportation, and utility. This chapter will focus on fuel cells for electric utility power generation, which represents the most likely large-scale use of fuel cells that will occur in this century.

A. Background

1. Historical

The fuel cell principle was first demonstrated in 1838 when Grove (1839) produced electricity from hydrogen and oxygen by operating a water electrolysis (electrochemical decomposition of water) cell in reverse. Four years later, Grove (1842) assembled a "gaseous voltaic battery" that used high-surface-area platinum electrodes, sulfuric acid electrolyte, and hydrogen and oxygen as the reactants. During his career Grove continued his investigations (Grove, 1874) into fuel cells and actually considered alternative fuels (to hydrogen), as well as the use of air as the oxidant.

During the 100 years from Grove's first fuel cell experiment to the 1930s, fuel cells were investigated by a number of scientists. The more notable investigations are described by Liebhafsky and Cairns (1968). Although no practical fuel cell evolved from these activities, many ideas were considered. In fact, most of the concepts that we find important in the 1980s were actually conceived prior to 1930, including high-surface-area electrode, gas diffusion electrodes, acid fuel cells, alkaline fuel cells, molten carbonate fuel cells, redox fuel cells, direct hydrocarbon fuel cells, direct coal fuel cells, and solid oxide fuel cells.

The efforts of Bacon (1954), starting in the mid-1930s and continuing into the 1960s, led the way to a practical fuel cell power plant. Bacon's fuel cell employed an alkaline electrolyte and metallic nickel electrodes to react hydrogen and oxygen electrochemically at 200°C and 600 psia. The Bacon cell concept was subsequently selected by Pratt and Whitney Aircraft as the basis for the Apollo spacecraft power plant. Modified to produce higher power densities and good reliability, the Bacon cell powered man's venture to the moon in the 1960s.

In addition to Bacon's efforts, other fuel cell activities were initiated in the 1950s. Allis-Chalmers explored a low-pressure alkaline fuel cell technology; General Electric (Grubb, 1957) developed an ion-exchange-membrane (now known as solid polymer electrolyte) fuel cell that was actually the first fuel cell in space, powering the Gemini spacecraft; Broers, working for the Central Technical Institute (TNO) in the Netherlands, pursued molten carbonate fuel cell technology. Led by the space efforts, interest in fuel cells grew explosively through the mid-1960s. Many potential applications were considered, including space, underwater, automotive, military, emergency power, and weather buoy. During this time period, almost every conceivable type of fuel cell was under investigation at one or more laboratories. At the peak of enthusiasm approximately 50 U.S. companies were involved. By the later 1960s, however, the problems inherent in fuel cell development had become apparent. This situation coupled with the reduction in emphasis on research and development in the United States, led to a rapid curtailment of fuel cell activity. And by 1970 only a few hardy organizations remained active in fuel cell development. In the United States, only United Technologies Corporation (formerly Pratt and Whitney Aircraft Division of United Aircraft) was making significant efforts aimed at consumer applications. The United Technologies Corporations (UTC) effort comprised two thrusts.

1. The TARGET (Team to Advance Research on Gas Energy Transformation) program to develop a fuel cell power plant for on-site

(residential/commercial) power generation was initiated in 1967 by the Pratt and Whitney Aircraft Division of United Aircraft and a consortium of gas companies.

2. An electric utility program was initiated in 1971 by United Aircraft, the Edison Electric Institute (EEI), and 10 electric utilities. In 1972 that program separated into (a) the FCG-1 program sponsored by nine utilities and United Aircraft and (b) the RP114 program sponsored by United Aircraft and the EEI. Responsibility for the RP114 program was transferred from the EEI to the Electric Power Research Institute (EPRI) in 1973. The FCG-1 program was intended to bring a first-generation, 26-MW fuel cell power plant into commercial service by the end of the decade. The RP114 program, in recognition of the somewhat limited applications of the first-generation machine, was aimed at seeking the technological advances needed to reduce cost and improve power plant flexibility.

The TARGET and FCG-1 programs were based on phosphoric acid fuel cell technology, while the RP114 effort was to explore alternative technologies and select the one best suited to future power plant application.

These programs provided the impetus for fuel cell development activities through 1975, when, partly in response to the energy crisis and partly due to renewed electric and gas utility interest in research and development, the federal government and the research arms of the utility industries initiated a comprehensive effort into fuel cell research and development. The results of this renewed fuel cell initiative have been impressive. A number of major companies are again actively involved. Furthermore, technological progress that was not forthcoming in the 1960s became rapid in the 1970s. As a result, fuel cell power plants are now on the threshold of commercial use.

2. Review of Nonelectric Utility Applications

Fuel cells are currently being developed for a number of diverse applications. In none of these can the fuel cell be considered an item of commerce. Although fully qualified as a spacecraft power system, the fuel cell has yet to achieve all of the goals required for widespread deployment as a commercial power plant. High efficiency (low fuel consumption), cleanliness (no inherent by-product or emissions), and lack of noise are the characteristics that drive most of the potential fuel cell applications. The major obstacle to deployment has been cost. The early fuel cell uses will, therefore, be those in which the premium for its advantageous characteristics and the tolerance to high capital cost are the greatest. Table I lists the most likely potential applications, together with the relative importance of the fuel cell's characteristics in each application.

Remote power applications in which the storage of fuel and/or oxidant is a problem are ideal for the fuel cell. This is especially true in the case of the longer-duration space and specialty undersea missions that are limited by the weight and volume of fuel and oxidant they can carry. For these applications, hydrogen–oxygen fuel cells are attrac-

TABLE I

Potential Fuel Cell Application[a]

	Importance of characteristic			
Application	High efficiency (low reactant weight/volume)	Low emissions	Quiet	Low cost
Remote power				
Space (1–20 kW)	1	2	3	3
Submarine (100–1000 kW)	1	2	3	2
Weather stations, buoys, communications, pipeline protection (0.1–100 kW)	1	3	3	2
Portable power				
Recreational (0.5–10 kW)	2	3	3	1
Military (0.5–10 kW)	1	1	2	2
Utility power				
Commercial/residential on-site (40–400 kW)	1	1	2	1
Industrial cogeneration (1–20 MW)	1	1	2	1
Electric utility generating stations (5–500 MW)	1	1	2	1
Other				
Emergency/standby (1–100 kW)	3	3	2	1
Transportation (20–4000 kW)	1	2	3	1

[a] Key: 1, major requirement; 2, important factor; 3, not a controlling characteristic.

tive, since they result in minimum weight/volume systems. The Gemini, Apollo, and Space Shuttle vehicles were all powered by hydrogen–oxygen fuel cells. The 1-kW Gemini power plant used the General Electric (GE) solid polymer electrolyte (SPE) technology, the 1.5-kW Apollo power plant was based on Pratt and Whitney Aircraft's version of the Bacon cell, and the larger 12-kW Space Shuttle fuel cell employs the UTC aqueous potassium hydroxide electrolyte. The fuel cell's record in space has been excellent during 7 Gemini flights, 12 Apollo flights, and the ongoing Shuttle activities. Both GE and UTC are developing advanced fuel cells for future space missions (McBryar, 1979).

Submarine applications have been restricted to a few specialty vehicles. This is likely due to the emergence of nuclear submarine fleets and the inability of conventional or fuel cell-powered submarines to compete. In the 1960s a substantial effort was conducted in Sweden to build a fuel cell propulsion system for submarines (Lindstrom, 1964), but it was terminated due to a series of engineering problems. United Technologies Corporation has developed and demonstrated a 60-kW (three 20-kW power plants) system for a deep-sea submergence vehicle (DSSV). It is possible that fuel cells will continue to serve specialty submarine applications; however, it is not a significant market opportunity.

A wide range of other remote applications that would be satisfied by fuel cells are characterized by a high energy-to-power ratio and the need for high efficiency and low environmental pollution. Navigational aids, automated weather stations, remote beacons, pipeline cathodic protection systems, and communication systems fall into this category. These applications range from milliwatts to about 100 W of power. In most cases they would use air as the oxidant and either methanol or hydrogen obtained from the decomposition of a metal hydride or methanol as the fuel. To date, a significant fuel cell market has not resulted from numerous investigations into this area, although recent Army interest in a family of 30- to 60-W fuel cell power plants based upon alkaline (George and Scozzofava, 1978) and SPE (Adlhart, 1978) cell technology could lead to a viable market.

Portable power applications overlap with air-breathing remote power applications. In our arbitrary classification system, portable power plants are defined as those in the 0.50- to 10-kW size range. They compete with such conventional generators as diesels and engine generator sets. For the recreational markets, the fuel cell's advantages are not sufficient to offset the much higher cost of the fuel cell as compared with engine generator sets. Thus, there is not yet any serious consideration of developing fuel cells for this use. Before such consideration becomes likely, fuel cell power plants will have to sell at $200/kW or less.

The U.S. Army Mobility Equipment Research and Development Command (MERADCOM) is developing a family of fuel cells for "forward-area" military use. In this service, reliability, silence, light weight, fuel efficiency, and nondetectability are of major importance. The prototype for this family is a 1.5-kW unit based upon the phosphoric acid technology and employing methanol (reformate) as its hydrogen source. The Army has contracted with UTC to conduct a 16-unit demonstration (Barthelemy, 1981). Under a parallel program, Energy Research Corporation is continuing with the development of 3- and 5-kW power plants (Abens *et al.*, 1981). The fuel cell will be cost competitive in this application at about $2000/kW and can meet special needs at as much as $4000/kW.

Utility applications can be classified into four categories:

1. those located on the customer side of the electric meter in commercial/residential situations (40- to 400-kW power requirements)
2. those located on the customer side in industrial cogeneration situations (1–20 MW),
3. those located on the electric utility side of the meter at substations (5–50 MW), and
4. those located on the electric side at central stations (100–500 MW).

The last three applications, which are generically similar, will be covered within the body of this chapter. The first, the so-called "on-site" application, has been pursued for about 15 years by a number of gas and gas–electric utilities.

The TARGET program was initiated in 1967 and in the late 1960s developed a 12.5-kW phosphoric acid fuel cell power plant, which demonstrated an electrical efficiency of 28%. In the early 1970s, 65 of these on-site power plants were tested at 35 sites in the United States, Japan, and Canada. Following these tests, the Gas Research Institute (GRI) and UTC developed and verified a larger 40-kW power plant, which incorporated heat recovery as well as several technical improvements. The 40-kW power plant is able to convert over 80% of the energy contained in the natural gas fuel to useful electrical and thermal energy.

In April 1980 the first of 48 field-test units entered into service at a Portland, Oregon, laundry. Related studies (Mientek, 1982) have indicated that the market size in terms of kilowatts sold is relatively insensitive in the 40- to 400-kW range. Above this range, the market diminishes significantly. Those same studies identified $2000–2500/kW as the threshold capital cost for this application.

The on-site power plants and the electric utility power plants share a great deal of common technology. In particular, the first-generation machines would utilize a very similar phosphoric acid fuel cell technology. For this reason, there is synergism among the various phosphoric acid fuel cell efforts. Technological accomplishments under the on-site program benefit the electric utility program and vice versa.

Other fuel cell uses that warrant mention are emergency power and transportation (propulsion) applications. Siemens AG has focused its attention on the development of a 7-kW fuel cell power plant for use as an emergency generator (Strasser, 1979). This application demands high reliability with low maintenance requirements. Efficiency and reactant type are not major concerns, since fuel consumption would be small or negligible during standby. Cost will be a major consideration; consequently, the Siemens approach uses H_2/O_2 reactants and low-cost alkaline electrolyte fuel cell technology. To date, experimental 7-kW systems have been tested successfully. The principal issues now relate to cost, life, and the European market for such a power plant. A well-defined U.S. market for emergency/standby power plants has not been identified, nor has a U.S. fuel cell been developed for such an application. Thus, the Siemens experience is being viewed with great interest.

In the United States transportation is the largest single energy user, accounting for approximately 50% of current petroleum consumption. This fact, coupled with the relatively slow development of battery-powered electric vehicles and the significant progress made by fuel cells, has stimulated interest in developing fuel cells for transportation (McCormick *et al.*, 1979). In transportation applications the fuel cell must compete with internal combustion engines and diesels that cost $10/kW and $50/kW, respectively. These applications will thus be driven by cost. Capital costs will likely control consumer markets (automobiles); life-cycle cost will be more important for the nonconsumer applications (buses, fleet vehicles, etc.).

The Los Alamos National Laboratory has conducted an ongoing evaluation of fuel cells for transportation (Huff, 1982), including passenger vehicles, city buses, delivery trucks, highway trucks and buses, and nonhighway heavy-duty transportation, such as locomotives and towboats. Power requirements range from 20 kW (60-kW peak) for the passenger vehicle to 400 kW for the city bus and 4000 kW for the push tug. Preliminary conclusions have been that the larger (power and size) applications are better suited to early use of fuel cells than are the smaller vehicles (passenger and delivery vans). The larger applications are nonconsumer, and users will consider life-cycle costs rather than first costs. Thus, fuel efficiency becomes a more important factor. Furthermore, larger fuel cell power plants (FCPP) will be less expensive in terms of dollars per kilowatt than the smaller plants due to economies of scale. For fuel cell power plants to be considered for any of these applications, they will have to use logistic fuels, i.e., gasoline, diesel, or perhaps methanol. Achievement of capital costs consistent with transportation applications will require technical advances that will likely take 15 to 20 years to implement and reduce to commercial practice.

In summary, the applications can be considered in two groups: specialty, where cost is not the major requirement, and consumer, where cost is all important. State-of-the-art fuel cell power plants are able to qualify now for many of the former applications. Given development funding and/or market incentives, fuel cells could be made available in a relatively short period of time for many of the specialty applications. However, fuel cell power plants are still marginally too expensive for the cost-sensitive applications. Table II indicates FCPP costs ranges that need to be achieved for such applications. This table suggests that

TABLE II

Cost-Sensitive Applications

Application	Cost range required for significant market (1982 $/kW)
Transportation	100–250
Emergency/standby	150–250
Recreational	200–300
Electric utility	700–1200
Industrial cogeneration	1000–1500
Commercial/residential on-site	2000–2500

the early commercial uses will be for electric utility, on-site, and industrial cogeneration applications.

B. Classification of Fuel Cell Technologies

Fuel cells are normally classified according to their fuel source, oxidant, temperature of operation, and electrolyte. Table III identifies some of these parameters. The number of combinations and permutations of the various parameters would seem to be very large (10 fuel sources × 3 oxidant types × 4 temperature ranges × 8 electrolytes). In fact, the number of practical combinations is relatively few for the following reasons:

1. Electrolytes are best suited to the following temperature ranges due to ionic conductivity and stability limitations: aqueous alkaline, SPE: low temperature (<120°C); phosphoric acid: intermediate temperature (~200°C); molten carbonate, molten alkaline: high temperature (~650°C); solid oxide: very high temperature (~1000°C).

TABLE III

Fuel Cell Parameters

Fuel source		Oxidant type	Temperature range	Electrolyte
Direct	Indirect[a]			
Hydrogen	Hydride	Oxygen (pure)	Low (≤120°C)	Aqueous acid
Hydrazine	Ammonia	Oxygen (air)	Intermediate	Phosphoric acid
Ammonia	Hydrocarbon	Hydrogen peroxide	(120–250°C)	Solid polymer
Hydrocarbon	Methanol		High (250–750°C)	electrolyte
Methanol	Ethanol		Very high	Aqueous alkaline
			(≥750°C)	Molten alkaline
				Aqueous carbonate
				Molten carbonate
				Solid oxide

[a] Indirect systems decompose the primary fuel external to the fuel cell to provide a secondary fuel that is fed to the fuel cell stacks. The secondary fuel is usually rich in hydrogen and may contain other products, such as water vapor and carbon oxides.

2. As discussed previously, the applications place constraints on the choice of fuel and oxidant.
3. There is a relationship between the fuel source (type) and the fuel cell temperature. In direct systems the fuel cell temperature must be high enough to result in reasonable electrochemical reaction rates. Hydrogen will react at any temperature; however, carbon monoxide (in coal gas) requires a higher temperature, and hydrocarbon fuels require an even higher temperature. Indirect systems often require that steam from the fuel cell (by-product water and heat) by cycled to the fuel-processing unit. This dictates a fuel cell temperature of at least 120°C to produce the steam.
4. Alkaline electrolytes react irreversibly with carbon oxides. Thus, carbon-containing fuels cannot be used with such electrolytes.

Table IV summarizes practical fuel cell combinations that are leading contenders for the various applications. As can be seen from this table,

TABLE IV

Practical Fuel Cell Combinations

Electrolyte	Temperature	Suitable application	Fuel	Oxidant
Aqueous alkaline	≤120°C	Space	H_2	O_2
		Submarine	H_2/N_2H_4	O_2/H_2O_2
		Other remote	Ind. hydride or methanol	Air
		Emergency	Ind. hydride/H_2	Air/O_2
Solid polymer electrolyte	≤120°C	Space	H_2	O_2
		Other remote	Ind. hydride	Air
		Low-power military	Ind. hydride	Air
		Emergency	Ind. hydride/H_2	Air/O_2
		Transportation	Ind. methanol	Air
Phosphoric acid	~200°C	High-power military	Ind. methanol or hydrocarbons	Air
		Utility power	Ind. methanol, hydrocarbons, or direct coal gas	Air
		Transportation	Ind. methanol	Air
Molten carbonate	~650°C	Utility power	Ind. methanol, hydrocarbons, or direct coal gas	Air
Solid oxide	~1000°C	Utility power	Any of the above	Air

a classification system based upon the electrolyte is very useful. Hence, it is commonly employed by fuel cell technologists. Furthermore, the state of development is also described by the electrolyte. For instance, aqueous alkaline, SPE, and phosphoric acid fuel cells are at an advanced stage of development, and demonstration is underway in at least one of the identified applications. The molten carbonate fuel cell is presently in transition from the single-cell stage to stack hardware. And the solid oxide fuel cell is slightly less developed than the molten carbonate.

C. Major Components of the Fuel Cell

The following are the three most important components of the individual fuel cell:

1. The anode (fuel electrode) must provide a common interface for the fuel and electrolyte, catalyze the fuel oxidation reaction, and conduct electrons from the reaction site to the external circuit (or to a current collector that, in turn, conducts the electrons to the external circuit).
2. The cathode (oxygen electrode) must provide a common interface for the oxygen and the electrolyte, catalyze the oxygen reduction reaction, and conduct electrons from the external circuit to the oxygen electrode reaction site.
3. The electrolyte must transport one of the ionic species involved in the fuel and oxygen electrode reactions while preventing the conduction of electrons (electron conduction in the electrolyte causes a short circuit). In addition, the role of gas separation in practical cells is usually provided by the electrolyte system, often by retaining the electrolyte in the pores of a matrix (or inert blotter). The capillary forces of the electrolyte within the pores allow the matrix to separate the gases, even under some pressure differential.

Other components may also be necessary to seal the cell, provide for gas compartments, and separate one cell from the next in the fuel cell stack.

A primary constraint imposed upon fuel cell components is that of material compatibility. The components must be stable in their respective environments for thousands (or tens of thousands) of hours. Thus,

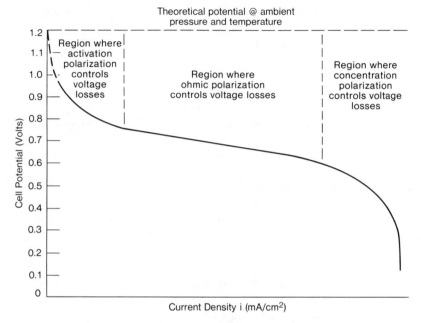

Fig. 3. Fuel cell performance or "polarization" curve.

the pacing challenge to the fuel cell technologist becomes that of developing the materials and components required for high-performance fuel cells in a hostile environment for extended periods of time.

D. Performance Characteristics

The performance of a fuel cell is represented by the current density versus cell potential (or "polarization") curve (Fig. 3). Whereas ideally a single hydrogen–oxygen fuel cell could produce 1.23-V dc at ambient conditions, in practice, fuel cells produce useful voltage outputs that are somewhat less than the ideal and that decrease with increasing load (current density). The losses or reductions in voltage from the ideal are referred to as polarization, as illustrated in Fig. 3. These losses include the following:

1. Activation polarization represents energy losses that are associated with the electrode reactions. Most chemical reactions in-

volve an energy barrier that must be overcome for the reactions to proceed. For electrochemical reactions, the activation energy lost in overcoming this barrier takes the form

$$\eta_{act} = a + b \ln i,$$

where η_{act} is activation polarization in millivolts, a and b are constants, and i is current density in milliamperes per square centimeter. Activation polarization is associated with each electrode independently and

$$\eta_{act(cell)} = \eta_{act(anode)} + \eta_{act(cathode)}.$$

2. Ohmic polarization represents the summation of all the ohmic losses within the cell, including electronic impedances through electrodes, contacts, and current collectors and ionic impedance through the electrolye. These losses follow Ohm's law:

$$\eta_{ohm} = iR,$$

where η_{ohm} is ohmic polarization in millivolts, i is current density in milliamperes per square centimeter, and R is total cell impedance in ohm square centimeters.

3. Concentration polarization represents the energy losses associated with mass transport. For instance, the performance of an electrode reaction may be inhibited by the inability of reactants to diffuse toward, or of products to diffuse away from, the reaction site. In fact, at some current, the limiting current density i_L, a situation will be reached wherein the current will be completely limited by the diffusion processes (see Fig. 3). Concentration polarization can be represented by

$$\eta_{conc} = (RT/nF) \ln[1 - (i/i_L)],$$

where η_{conc} is concentration polarization in millivolts, R a gas constant, T temperature in Kelvin, n the number of electrons, F Faraday's constant, i current density in milliamperes per square centimeter, and i_L limiting current density in milliamperes per square centimeter. Concentration polarization occurs independently at either electrode. Thus, for the total cell

$$\eta_{conc(cell)} = \eta_{conc(anode)} + \eta_{conc(cathode)}.$$

The net result of these polarizations is that practical fuel cells produce between 0.5 and 0.9 V dc at currents of 100 to 400 mA/cm^2 of cell area.

Fuel cell performance can be increased by increasing cell temperature and reactant partial pressure. For any fuel cell, the trade-off always exists between achieving higher performance by operating at higher temperature or pressure and confronting the materials and hardware problems imposed by the more severe conditions.

E. Fuel Cell Technologies and Electric Utility Power Generation

In the near term, electric utility fuel cell power plants will be based upon the phosphoric acid fuel cell (PAFC) technology. This technology has several advantages.

1. It is at a mature stage of commercial development.
2. It naturally rejects the carbon dioxide present in either reformed hydrocarbon fuels or coal gas.
3. It can operate at a temperature of 200°C, where the carbon monoxide contained in the fuel stream is not a serious poison and where there is efficient thermal integration between the fuel cell and the fuel processor.
4. Waste heat of 200°C can be used for cogeneration applications (although these are somewhat limited by the relatively low temperature).
5. Materials of construction other than the platinum catalyst are low cost (and only very small quantities of platinum are required).
6. The cell can operate over the broad range of fuel gas compositions and external environments that might be expected in utility applications.

The aqueous alkaline technology is not applicable because of its incompatibility with carbonaceous fuels, and the solid polymer electrolyte technology is not well suited to operation at temperatures above 120°C because of its tendency to dry out if exposed to low-humidity gases (air) at elevated temperature.

In the longer term, the molten carbonate fuel cell (MCFC) is a candidate for utility use. Its advantages include the following.

1. It is compatible with carbonaceous fuels and, in fact, requires the use of such a fuel to provide a source of carbon dioxide—a reactant at the cathode.

2. It operates at a temperature of 650°C, where carbon monoxide is utilized within the fuel cell and is, in effect, not distinguishable from hydrogen in its behavior within the cell.
3. At 650°C, thermal integration with a variety of fuel processors is possible, and waste heat can be used for a wide range of industrial cogeneration applications.
4. Expensive catalysts are not required.

Another, longer-term alternative is the solid oxide fuel cell (SOFC) that operates at 1000°C. It enjoys the same advantages as the MCFC. Its primary limitation has resulted from the need to develop materials with adequate performance and life at that very high temperature.

II. PHOSPHORIC ACID FUEL CELLS

A. Description

A section of a typical phosphoric acid fuel cell (PAFC) is depicted in Fig. 4. It consists of the following components:

1. *A carbon or graphite separator–current collector plate* that separates hydrogen from the air of the adjacent cell (in a multicell stack) and also provides the electrical series connection between cells: This plate must be impermeable to hydrogen and air (oxygen), a good electronic conductor, and stable to both fuel and air environments in the presence of concentrated, 200°C phosphoric acid.
2. *Anode current collector ribs* that conduct the electrons from the anode to the separator plate: The ribbed configuration provides gas passages for hydrogen distribution to the anode. These carbon or graphite ribs are made a part of either the separator current collector plate or the anode. In either event, they must have good electronic conductivity and be stable to the fuel environment. If made a part of the separator plate, they would be fabricated of dense, impermeable material, but if they were part of the anode, they could be porous. The latter configuration would appear to have two advantages over the ribbed separator–plate: the ribs can be formed in a continuous manufacturing process; and if porous, the ribs can also be used to store phosphoric acid, thereby increasing endurance.

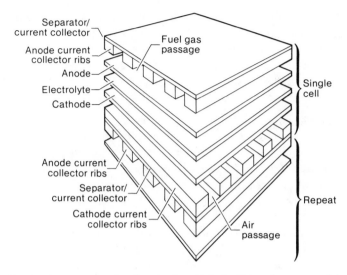

Fig. 4.   Components of a phosphoric acid fuel cell (not drawn to scale).

3. *An anode* (fuel electrode) that consists of a porous, graphite substrate with the surface adjacent to the electrolyte treated with a platinum or platinum alloy catalyst: Practical anodes will contain only a fraction of a milligram of noble metal for a square centimeter of cell area (0.25 mg/cm^2 is a typical state-of-the-art loading). Teflon is usually employed as a catalyst binder.

4. *An electrolyte matrix* that retains the concentrated phosphoric acid: The state-of-the-art matrices are fabricated from silicon carbide powder and Teflon binder and range between 0.01 and 0.03 cm in thickness. The electrolyte matrix should have minimum ionic resistance while separating the fuel and oxidant gas streams. To accomplish this separation in a practical power plant, the matrix should have pores or capillaries that are sufficiently small so that when filled with electrolyte, they can withstand pressure differentials of 0.06 atm or more without gas "blow through."

5. *A cathode* (oxygen electrode) that is similar to the anode but may use a modified noble metal catalyst and an increased catalyst loading (usually 0.5 mg/cm^2) to enhance the oxygen-reduction kinetics: In addition, the cathode is made hydrophobic by increasing the Teflon content. This prevents flooding of the electrode and provides for adequate oxygen diffusion to the active sites.

6. *Cathode current collector ribs* that are also virtually identical to the anode ribs: The only differences would be in the height and width of the ribs that control the mass transport characteristics in the gas passage and the contact resistance between components. In a stack, cathode ribs would be perpendicular to the anode ribs to simplify the location of external manifolds.

The individual fuel cells are stacked in series, as shown in Fig. 5, to produce the desired power and voltage output. For instance, the 4.5-MW power plant being installed in New York contains 20 stacks, each having nearly 500 cells of 3.7-ft^2 area each. Fuel and air supply and exhaust manifolds are then connected along the respective sides of the stacks. Both cell seals and manifold seals are critical to proper operation. The cell seals are located between the separator plate and the electrolyte matrix to prevent overboard leakage of fuel or air. The manifold seals are located between the respective sides of the stack and the manifolds. The cell sealing is usually accomplished by a "wet seal"; that is, the surface tension of the phosphoric acid wetting the component surfaces is sufficient to provide sealing. The substrate is densified in the seal area to improve sealing. The manifold seal is accomplished with a Teflon caulking.

To simplify sealing at high pressures, the stacks are housed in a pressure container. This allows the container to be pressurized to operating pressure and minimizes the differential pressure across the seals.

B. Performance Characteristics

Figure 6 indicates typical performance characteristics (Fickett, 1977) for a PAFC at the start of a test. Important PAFC performance characteristics include the following.

1. Present PAFC performance is almost entirely determined by the cathode activation polarization. That is, concentration polarization, anode activation polarization, and ohmic losses are small compared to the activation losses exhibited at the cathode. For instance, almost 300 mV (of a theoretical 1230 mV) are lost at the cathode at very low current densities.
2. Cathode current density (at fixed potential) is proportional to oxygen partial pressure. Thus, increasing the system pressure (and, hence, the oxygen partial pressure) can have a dramatic effect on performance.

Fig. 5. Fuel cell stack. (Courtesy of United Technologies.)

3. Cathode current density (at fixed potential and pressure) has been shown (Kunz and Gruver, 1975) to be essentially proportional to catalyst loading (for loadings of 1 mg/cm² and less). Above this loading, current density gains diminish with further increases in catalyst.

4. Utility PAFC power plants will derive hydrogen by the conversion of a wide variety of primary fuels such as methane (natural gas), petroleum products (such as naphtha), coal liquids (such as

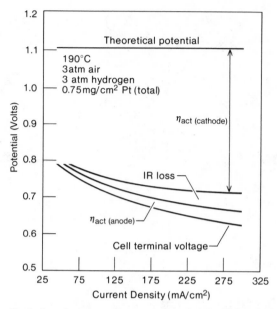

Fig. 6. Typical performance characteristics of phosphoric acid fuel cells.

methanol), or coal gases. The conversion process will result in a hydrogen-rich fuel gas containing CO_2 and CO. The CO_2 is inert and acts as a diluent; the CO, however, is a potential fuel cell poison. The low anode polarization portrayed in Fig. 6 can be achieved only under relatively low CO concentrations (Stonehart and Baris, 1980). Figure 7 indicates the CO concentrations and temperatures at which the anode performance shown in Fig. 6 is attainable.

5. Within practical power plant operating conditions, cell temperature has little impact on performance (other than the CO tolerance effect discussed in point 4). For a given pressure, an increase in temperature increases the phosphoric acid concentration. Thus, any gain due to the effect of temperature on kinetics is offset by the reduced intrinsic activity of the more concentrated acid. However, for a practical power plant, it is not possible to decouple pressure and temperature, since the waste heat from the fuel cell power section is used to raise steam for the fuel processor. The fuel cell temperature must be high enough to produce steam

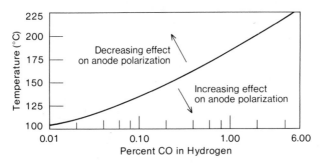

Fig. 7. Relationship of temperature and carbon monoxide concentration to phosphoric acid fuel cell anode polarization.

at a pressure sufficient for injection into the fuel processor. This dictates certain pressure–temperature combinations, i.e., 3 atm–190°C, 6 atm–205°C, 8 atm–210°C. These combinations allow sufficient temperature to overcome thermal losses and heat exchanger "pinch" points while still producing steam at the required pressure.

Given these points, the major opportunities for improving PAFC performance include the following.

1. Developing catalysts to improve the cathode kinetics: To this end, results (Ross, 1980) have shown significant performance improvement with intermetallic catalysts (Fig. 8). For example, at a cathode potential of 0.7 V dc, the use of Pt-V allows the current density to be approximately doubled. Thus, the development of stable alloy catalysts may be an important route to better performance.

2. Increasing the oxygen partial pressure by operating the system at higher overall pressure: At fixed cathode potential, the current density will be roughly proportional to oxygen partial pressure or approximately proportional to system total pressure (if water partial pressure is neglected). This requires increasing cell temperature, since the pressure and temperature are coupled.

Figure 9 projects initial PAFC cell performances that might be expected with conventional and improved catalysts at several pressure–temperature conditions.

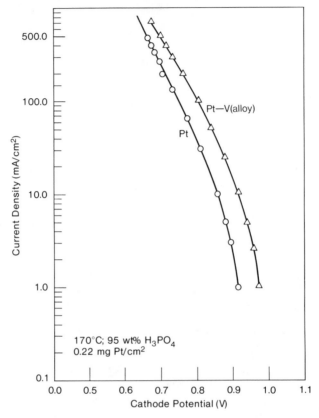

Fig. 8. Performance of Pt–V intermetallic alloy catalyst versus Pt. [From Ross (1980).]

C. Endurance Characteristics

Figure 10 depicts typical PAFC performance decay as a function of operating time at a constant current density. The interesting feature of this figure is the linear relationship of voltage decay with log (time) following the initial thousand hours of operation. This decay results from the tendency of the platinum catalyst to lose surface area and, hence, intrinsic activity. This loss is due to three mechanisms.

1. The carbon substrate slowly oxidizes at the fuel cell cathode, resulting in both a reduction of surface area and a spalling of platinum.

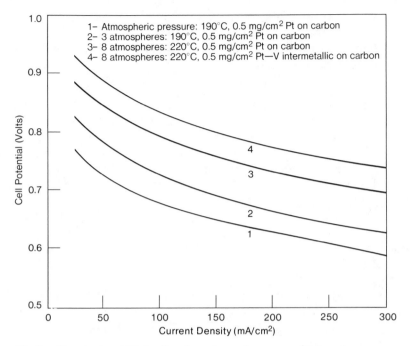

Fig. 9. Phosphoric acid fuel cell performance under a range of temperature–pressure conditions.

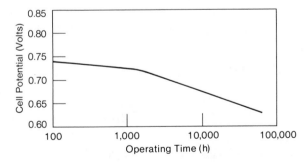

Fig. 10. Typical performance decay for phosphoric acid fuel cells.

2. With time, the platinum crystallites migrate on the carbon surface and aggregate to form larger crystallites (smaller surface area).
3. Platinum oxide tends to be slightly soluble in the acid and undergoes dissolution–reprecipitation, resulting in a slow growth in crystallite size with time.

The net result of these three mechanisms is a loss of cathode catalyst surface area that is a function of log (time). These mechanisms are becoming understood, as are the parameters that control the decay rates (Stonehart and MacDonald, 1981). Cathode potential, temperature, acid concentration (water content), and oxygen partial pressure are all important parameters. Even more important, however, are the characteristics of the carbon (graphite) surface and the platinum alloy composition. Carbon oxidation as well as platinum migration and dissolution can be markedly reduced by careful attention to the nature of the materials. Thus, the decay rates shown in Fig. 10 will decrease with a change in operating parameters as well as with subtle material changes.

It is important that the end-of-life performance can be modeled and projected from relatively short-term cell data; there are now sufficient data to verify that projections such as Fig. 10 do indeed portray the performance characteristics of the better cells and cell stacks.

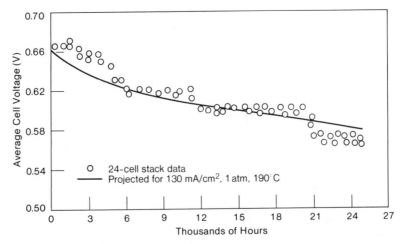

Fig. 11. Endurance test results of a 24-cell phosphoric acid fuel cell stack. [From Mientek (1982).]

There are other potential decay and failure modes that can cause cells to lose performance and fail. These other mechanisms, which would result in deviations from Fig. 10, include resistance increase and/or gas crossover due to insufficient electrolyte, concentration polarization at the cathode due to inadequate hydrophobicity (Teflon) resulting in flooding (loss of cathode reaction area), loss of catalyst activity due to poisoning of the cathode or anode, and external gas leakage. All of these problems can and do occur; however, they are not considered to be inherent in the PAFC technology and can be avoided through proper attention to design and fabrication.

Figure 11 shows the performance of a 24-cell stack that has operated for more than 24,000 hours. This stack verifies the ability to achieve or better the endurance characteristics depicted in Fig. 10.

III. MOLTEN CARBONATE FUEL CELLS

A. Description

The molten carbonate fuel cell (MCFC) structure is geometrically similar to that of the PAFC structure as shown in Fig. 4. The materials that are used are, however, very different from those used in the PAFC. Referring to Fig. 4, the MCFC consists of the following components.

1. *A separator–current collector plate* that, like its PAFC counterpart, must be impermeable to hydrogen and oxygen, a good electronic conductor, and stable to fuel and air environments in the presence of 650°C carbonate salts: No single material has been found that is completely satisfactory for this duty—those satisfactory in the fuel environment are undesirable in the oxidizing environment and vice versa. The present solution is to use a bimetallic sheet such as nickel (fuel side) clad with stainless steel (oxidant side).

2. *An anode current collector* that conducts the electrons from the anode to the separator plate and provides passage for fuel flow: In some MCFC configurations, this function is provided by ribbing or folding the separator plate. In other configurations, the current collector ribs are formed into the porous electrodes, as discussed under PAFCs. The current collector must have good electronic

conductivity and be stable to the fuel environment. Nickel is a satisfactory material, although copper is a possibility if it can be treated to reduce sintering.

3. *An anode* that consists of a porous nickel treated with a refractory oxide to reduce sintering: At the 650°C temperature, no other catalyst is required.

4. *An electrolyte system* comprising a mixture of lithium–potassium carbonate and inert power (presently lithium aluminate): This mixture forms a paste when molten. When cooled, however, the mixture freezes to form a rigid "tilelike" structure. This electrolyte system presents major challenges to MCFC development. It must have minimum ionic resistance while separating the fuel and oxidant gases. In addition, it must be electronically insulating. A historical problem with MCFCs is the tendency for the electrolyte system to crack due to stresses induced during the cooling of the hardware below the carbonate freezing point. If these cracks do not "heal" upon subsequent melting, the fuel and oxidant gases mix, causing premature failure.

5. *A cathode* that is similar to the anode except that it presently uses nickel oxide (doped with lithium to impart electronic conductivity): Since hydrophobic agents, such as the Teflon used in PAFCs, are not stable at 650°C, cathode flooding is prevented by careful attention to the pore (capillary) sizes in the anode, the electrolyte system, and the cathode. The cathode pore size must be larger than the other two to prevent flooding. (The electrolyte system must, of course, have the smallest pores to retain the molen carbonate.) Alternative cathode materials need to be developed, since nickel oxide is not stable for long times at the higher operating pressures that will likely be required for good performance.

6. *A cathode current collector* that has similar requirements and configurational options as the anode current collector: Since nickel is thermodynamically unstable, material options include corrosion-resistant stainless steel.

As with PAFCs, individual cells are stacked in series to result in a cell stack of the required power and voltage output. Cell and manifold sealing is accomplished by wet seals and the use of an inert caulking (again similar to PAFC with different materials). When operating at

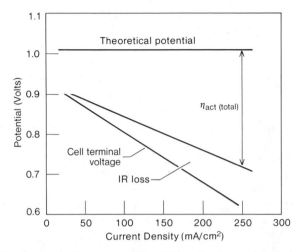

Fig. 12. Typical performance characteristics of molten carbonate fuel cells.

pressure, the stacks will be contained in a pressure vessel to minimize the pressure differential across the cell and stack seals.

B. Performance Characteristics

Figure 12 illustrates a typical MCFC performance at the start of test. The performance of MCFCs tends to be slightly better than that of PAFCs primarily because of the improved behavior of the air cathode. Performance considerations include the following:

1. The performance losses in practical MCFCs are distributed among a variety of polarization components, including both cathode and anode activation polarization, as well as concentration polarization and ohmic polarization. The cathodic activation and ohmic polarization are probably the largest contributors.
2. Ohmic losses due to electrolyte and contact resistance can be severe; thus, attention must be given to developing thin electrolyte structures (≤ 0.05 cm) and maintaining good pressure contacts. Total cell ohmic losses should be maintained below 40 mV at 100 mA/cm^2.
3. The MCFC involves the reactions of the CO_2 at the cathode, $CO_2 + \frac{1}{2}O_2 + 2e \rightarrow CO_3^{2-}$ and its release at the anode, $CO_3^{2-} + H_2 \rightarrow$

$CO_2 + H_2O + 2e$. Thus in a practical MCFC system, CO_2 must be recycled from the anode exhaust to cathode inlet. The kinetics of the cathode are extremely complex and not completely understood but involve peroxides or superoxides as well as carbonates. Therefore, although thermodynamics would suggest that the ratio of CO_2 to O_2 in the oxidant feed should be 2 to 1, the kinetics may optimize at a different ratio.

4. Current density will increase with increased operating pressure. The precise theoretical relationship is not well understood; however, empirical data for the pressure effect are shown in Fig. 13. System considerations have generally caused MCFC cost and efficiency to optimize at between 6 and 10 atm.

5. Any CO in the fuel gas will undergo the water/gas shift reaction $(CO + H_2O \rightarrow CO_2 + H_2)$ at 650°C cell temperature. The equilibrium constant for this reaction is a function of temperature; the theoretical potential decreases with increasing temperature. However, all other performance aspects improve with increasing temperature. The net effect is improved cell voltage with increased temperature (Fig. 14). The relatively small effect of temperatures above 650°C provides little incentive for operating above this temperature and thus encountering the additional corrosion problems that result.

6. In addition to the water/gas shift reaction, other reactions can occur at the anode: the methanation reaction [CO (or CO_2) + $3H_2 \rightarrow CH_4 + H_2O$] and carbonization reactions [$2CO \rightarrow C +$

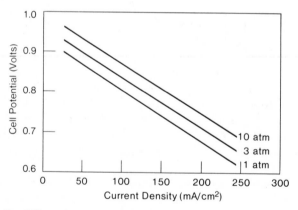

Fig. 13.   Effect of pressure on molten carbonate fuel cell performance.

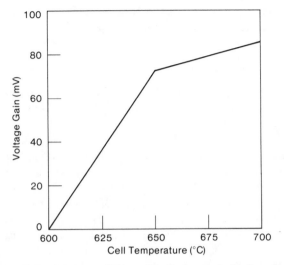

Fig. 14.   Effect of temperature on molten carbonate fuel cell performance.

CO_2] may take place, removing available hydrogen (or CO) from the anode stream and reducing performance. Thus, the fuel composition and flow rate (utilization) can have significant impact on performance as well as on other system characteristics.

7. Because CO reacts via the water/gas shift at the MCFC anode, it is a desirable component of the fuel gas (virtually indistinguishable from H_2 in its effect on the fuel cell, rather than being a poison as in the PAFC case). The most obvious MCFC poison is H_2S or COS. Concentrations as low as 1 ppm can have a deleterious effect. Thus, MCFC systems must exhaustively scrub sulfur-containing species before they enter the cell. Surprisingly, this process does not inflict a severe cost or efficiency penalty upon the power plant. Reducing the sulfur content of a typical coal gas to ≤ 1 ppm may add \$40/kW to the capital cost and reduce the efficiency by 1% (Lorton, 1980).

An excellent discussion of the various performance parameters is contained in the *Handbook of Fuel Cell Performance* (Benjamin *et al.*, 1980).

In summary, state-of-the-art MCFC performance is good, and future gains will be derived from reduced ohmic resistance, slightly improved anode and cathode transport characteristics, increased operating pressure, and careful attention to the total system to optimize fuel and oxidant composition and utilization.

C. Endurance Characteristics

Figure 15 shows the best life-test data obtained on a small single cell. The performance decrease with time tends to be associated with a corresponding ohmic increase.

Loss of electrolyte is one of the major modes of MCFC performance decay. This loss can occur either through the reaction of the electrolyte with other cell and stack components or through evaporation into the fuel or air exhaust streams. The cell shown in Fig. 15 had electrolyte added to it throughout the test. As electrolyte is lost from the cell to the point that unfilled pores or cracks exist in the electrolyte system, reactant gases will mix, resulting in cell overheating and failure. Practical MCFCs will require adequate electrolyte inventory within the cell hardware for 40,000 hours. This inventory will be significantly reduced at higher operating pressure, since the evaporative losses will be reduced as an inverse function of pressure.

In addition to electrolyte loss, performance decay can be caused by the corrosion of the various cell components. Even stainless steel may be expected to corrode within MCFCs.

Another potential degradation mechanism is the tendency of the lithium aluminate filler in the electrolyte system to undergo phase changes as a function of temperature and time. The phase changes result in volume (pore size) changes leading to a loss in the system's ability to retain electrolyte. Alternative filler materials are being inves-

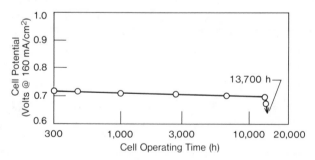

Fig. 15. Performance versus time for longest-lived molten carbonate fuel cell. [From Houghtby *et al.* (1978).]

tigated. However, low-cost materials that are stable in this system are difficult to find.

The most serious endurance problem for state-of-the-art MCFCs is not long-term decay but rather a precipitous failure caused by reactant mixing through open (unfilled) cracks in the electrolyte system. This problem usually results from a thermal cycle—that is, a cell cooldown to below the electrolyte's melting point. Thermal stresses that develop during a shutdown (cooldown) cause the electrolyte system to crack. If this crack cannot heal itself upon the next heatup (melting), then the reactants will mix and the cell will fail. Major research and development efforts addressing both material and configurational solutions appear to offer a range of solutions.

The longest endurance demonstrated by a multicell stack is 3000 hours (Bett *et al.* (1980). This stack composed of ten 1-ft^2 cells also completed thermal cycles without failure, confirming at least limited success in resolving this problem.

IV. SOLID OXIDE FUEL CELLS

A. Description

The solid oxide fuel cell (SOFC) employs a "tubular" rather than a "planar" cell configuration. Using this tubular cell, depicted in Fig. 16, avoids the problem of sealing the cell edges, a necessity in planar designs. Such edge seals are considered impractical due to the lack of nonporous, insulating gasket materials for use at 1000°C. As Fig. 16 illustrates, the state-of-the-art SOFC consists of five basic components: a porous support tube, a fuel electrode, a solid electrolyte, an air electrode, and an electronically conducting interconnection.

Practical considerations in developing the individual components include the following.

1. *Porous support tubes* are required to provide a mechanically strong structure. The support tube must allow access of the oxygen to the air electrode; thus, a proper trade-off between strength and porosity is important. It is also necessary that the thermal expansion characteristics of the support tube match those of the other components. These considerations, coupled with cost, have led to the selection of calcia-stabilized zirconia as the material

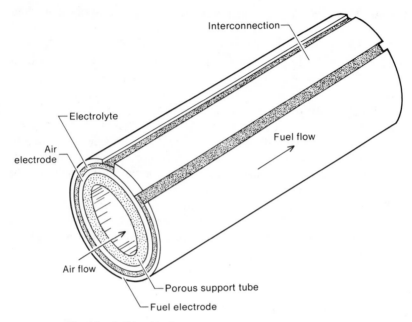

Fig. 16.   Solid oxide fuel cell. (Courtesy of Westinghouse.)

presently in use in the Westinghouse program. The Westinghouse requirements include 25-vol% open porosity, 5000-psi tensile strength, and 10.5×10^{-6}-m/m °C thermal expansion coefficient.

2. *The air electrode (cathode)* must conduct electrons, withstand the oxidizing environment, allow oxygen to reach the electrolyte interface, adhere to the electrolyte, thermally cycle without cracking, and catalyze the reduction of oxygen to oxide ions. Only electron-conducting metal oxides are able to satisfy these requirements. Candidate materials include tin-doped indium oxide, doped lanthanum nickel oxide, and doped lanthanum manganite. To enhance performance, a catalyst such as praesodymium oxide can be incorporated into the electrode–electrolyte interface. A concern is the mechanical adhesion between the air electrode and the electrolyte. Spalling or flaking during thermal cycling has been a problem.

3. *The fuel electrode (anode)* must be electronically conductive, must not crack or lose conductivity after a thermal cycle, must allow fuel gas to reach the electrolyte interface, and must catalyze

the fuel oxidation reaction. The use of a porous nickel–zirconia cermet ensures a thermal expansion match and reasonable conductivity. The zirconia also provides a bond to the porous support and mitigates the sintering of the nickel catalyst.

4. *The solid electrolyte* must be an absolute gas barrier, provide good ionic conductivity, but be an electronic insulator. (This last point is extremely important and will be developed later in the context of lower-temperature solid oxide electrolytes.) The SOFC electrolyte must tolerate thermal cycling without loss in integrity.

Figure 17 plots the ionic conductivity of several oxide conductors. As can be seen, the best conductivities are obtained with 10% yttria-stabilized zirconia (about 10^{-1} mho/cm at 1000°C), and 4% yttria–4% ytterbia-stabilized zirconia (about 2×10^{-1} mho/cm at 1000°C). In addition, these electrolytes have negligible electronic conductivity and are stable in the fuel cell environment.

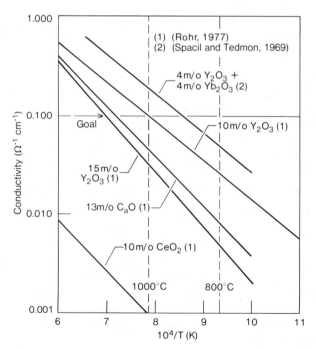

Fig. 17. Conductivity of various stabilized zirconias.

Brown, Boveri, and Cie AG has demonstrated four years of successful operation with cells containing yttria–ytterbia-doped zirconia (Rohr, 1977).

5. *The interconnection* between cells represents one of the major challenges in SOFC development. The interconnection must have high-electron conductivity (and negligible ionic conductivity), gas tightness, low volatility, and a thermal expansion match with other components, and it must be chemically compatible with both oxidizing and reducing environments. Compatibility with an oxidizing environment dictates the use of an oxide (nothing else is stable in air at 1000°C); yet most oxides are thermodynamically unstable in a reducing atmosphere, and only a few offer electronic conductivity. The recent Westinghouse effort has focused on this problem and has identified modified lanthanum chromites as suitable cell interconnection materials.

B. Performance Characteristics

Figure 18 depicts typical single-cell performance on hydrogen and air and hydrogen–carbon monoxide and air. Performance considerations include the following.

1. Electrode (cathode and anode) polarization losses are small compared to the those of either the PAFC or the MCFC.
2. Ohmic resistance losses are high. Of special note is the resistance within the cathode. This resistance, which is negligible in other fuel cells, is a significant factor in the case of the SOFC.
3. The SOFC is tolerant of a wide range of fuel gases and has been shown to tolerate up to 50 ppm of sulfur with nickel cermet anodes and 200 ppm of sulfur with cobalt cermet anodes.

A patent issued to Hitachi Ltd. in 1971 (Maki *et al.* 1971) describes the development of a solid oxide electrolyte with sufficient conductivity at 700°C to be useful in practical fuel cells. Subsequently, there was considerable activity aimed at developing fuel cells based on solid-oxide materials that conduct at temperatures of 700 to 800°C. Operation at these lower temperatures would circumvent the materials and lifetime problems inherent at 1000°C temperature range. Ceria doped with gadolinia, yttria, and calcia are typical of oxides that exhibit high conductivity at 700 to 800°C.

Further investigations (Kudo and Obayashi, 1976; Ross and Ben-

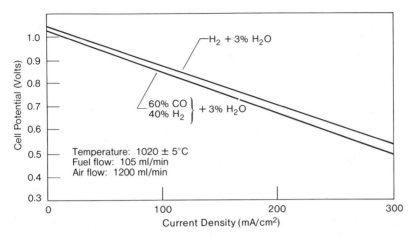

Fig. 18. State-of-the-art performance of solid oxide fuel cells. [From Isenberg (1982).]

jamin, 1977) determined that these solid oxides become mixed ionic–electronic conductors at low oxygen partial pressures (on the fuel side). This, in effect, creates an electronic short across the electrolyte; and although seemingly small, this electronic short drastically reduces the achievable cell efficiency (to less than 40%), which, in turn, limits the total power plant efficiency to less than 35%. Ross and Benjamin (1977) and Kudo and Obayashi (1976) provide excellent treatments of the mechanisms that are responsible for this electronic conduction and loss of efficiency. This literature should be carefully reviewed before attempting to use low-temperature solid oxide electrolytes.

C. Endurance Characteristics

The endurance capability of SOFCs has steadily improved. Single-cell lifetimes of 34,000 hours have been achieved at a constant output of 120 mA/cm^2 and 0.7 V dc (Rohr, 1977).

More than 3000 hours of operation have been achieved on a 10-cell stack (Isenberg, 1980). After 2000 hours of operation at 400 mA/cm^2 on hydrogen–air, stack voltage had dropped 6%. This stack subsequently operated stably for an additional 3000 hours at 150 mA/cm^2 on simulated coal gas. Both tests involved 11 cooldown and reheat cycles; the findings supported some confidence that cells and stacks can tolerate thermal cycling without performance deterioration.

V. ELECTRIC UTILITY FUEL CELL SYSTEMS

A. Electric Utility Fuel Cell Power Plant Characteristics

An electric utility fuel cell power plant will include the three principal subsystems described in Fig. 2: the power section, the fuel processor, and the power conditioner. The power section receives a hydrogen-rich gas stream from the fuel processor and delivers power at 2000 to 3000 V dc to the power conditioner. The fuel processor converts a logistic (available) utility fuel such as natural gas, naphtha, or coal gas to the hydrogen-rich gas received by the power section. The power conditioner converts the dc power to ac power compatible with the utility grid.

In assessing the characteristics of the utility fuel cell, this entire power plant must be considered. All references to fuel cell power plant characteristics—whether environmental, efficiency, cost, or other—must include this total process and all of the equipment required to convert the logistic utility fuel to ac power at the grid.

Fuel cell power plants have several attributes not found in thermal power generators.

1. High efficiency: A level of 41% (based on fuel higher heating value) is projected for initial units, with a potential for over 50% in the future. Fuel cell efficiencies are relatively high over a wide range of load and down to sizes of 10 MW or less. Figure 19 compares the efficiency characteristics of the fuel cell with those of other generators.
2. Environmental compatibility: NO_x, SO_x, and particulate emission levels are lower than any projected requirements. Fuel cell plants are also water conservative and quiet.
3. Modular design: Because fuel cells will be factory produced with a short lead time and because performance is virtually independent of plant size, they can be used to increase utility system capacity by small increments in response to demand growth. Capital outlay is thereby reduced, and the effect of interest charges during construction is minimized.
4. Versatility: Fuel-processing and power-conditioning options provide flexibility in both the type of fuel used and the power introduced into the bus. For example, the power conditioner can be used to control real and reactive power independently.

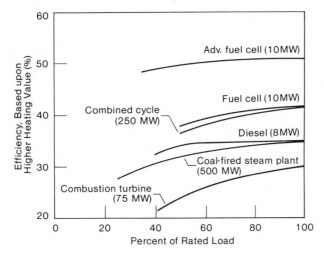

Fig. 19. Efficiency of electric generation options.

5. Dual energy use: Waste heat can easily be extracted from a fuel cell power plant without degrading the other performance characteristics. Operating in a dual energy use (electric and thermal) mode, a fuel cell can be 80% efficient.

Table VA summarizes the projected characteristics of both the first generation phosphoric acid fuel cell (PAFC) power plant and future (advanced acid, molten carbonate, or solid oxide power plants) operating on liquid or gaseous fuels. Table VB summarizes these same characteristics for power plants that are integrated with a coal gasifier.

B. Electric Utility Applications

Figure 20 illustrates possible fuel cell locations on the utility grid, which include the following.

1. *Customer side of the electric meter, fueled by natural gas, and with waste heat recovery:* These applications range from the 40- to 400-kW on-site units in residential/commercial locations (of primary interest to the gas and gas–electric utilities) to 1- to 20-MW industrial cogeneration power plants (of interest to both gas and electric utilities). The primary fuel for these units would be

TABLE VA

Projected Fuel Cell Power Plant Characteristics (with Liquid–Gaseous Fuel)[a]

Characteristic	First generation (phosphoric acid)	Future (advanced acid, molten carbonate, or solid oxide)
Available for order (year)	1986	1990–1995
Module size (MW)	~10	~10
Capital cost ($/kW)[b]	660	640
Lead time (years)	2	2
Operation/Maintenance cost		
(mills/kW h)	5	5
Life (years)[c]	30	30
Efficiency: electric only (%)		
Full load	41	50
Half-load	38	50
Efficiency: dual energy (%)	80	80
Emissions (lb/10^6 Btu)		
SO_x	3×10^{-5}	3×10^{-5}
NO_x	2×10^{-2}	2×10^{-2}
Particulates	3×10^{-6}	3×10^{-6}
Water requirement (95°F ambient)	None	None
Noise (dBa at 100 ft)	≤55	≤55
Start-up time (hours)		
Cold	≤6	≤4
Hot	≤1	≤1
Transient response (seconds from minimum to maximum load)	10	15
Power factor correction (degrees arc)	−90 to +90	−90 to +90

[a] Fuel options: natural gas, naphtha, clean coal, alcohols.

[b] Including interest during construction and installation; December 1981 dollars; assumes production of 500 MW/year.

[c] Book life with cell stack replacement every 40,000–60,000 hours.

natural gas, although opportunistic fuels, such as an individual by-product gas, might be used in a specific situation.

2. *Multimegawatt fuel cell power plants located at transmission substations:* These power plants could range from 5- to 50 MW in size. Because these dispersed power plants would be located close to the users of energy (both electric and thermal), waste heat recovery will likely be an important consideration. The dis-

TABLE VB

Projected Fuel Cell Power Plant Characteristics (Integrated with Coal Gasifier)

Characteristic	First generation (phosphoric acid)	Future (advanced acid, molten carbonate, or solid oxide)
Fuel option	Coal	Coal
Available for order (year)	1988	1995
Module size (MW)	50	50–500
Capital cost ($/kW)[a]	1440	1000
Lead time (years)	4	4
Operation/Maintenance cost (mills/kW h)	5	8
Life (years)[b]	30	30
Efficiency: electric only (%)	34	50
Emissions (lb/10^6 Btu)[c]		
SO_x	TBD	TBD
NO_x	TBD	TBD
Particulates	TBD	TBD
Water requirement (95°F ambient)	TBD	TBD
Noise (dBa at 100 ft)	55	55
Start-up time (hours)		
Cold	≤6	≤4
Hot	≤1	≤1
Transient response (seconds from minimum to maximum load)	10	15
Power factor correction (degrees arc)	−90 to +90	−90 to +90

[a] Including interest during construction and installation; December 1981 dollars; includes coal gasifier.

[b] Book life with cell stack replacement every 40,000–60,000 hours.

[c] TBD means to be determined.

persed nature of the siting dictates the use of relatively clean, easily transported fuels, i.e., liquids or gases.

3. *Central station fuel cell power plants:* These would typically be located at a distance from the ultimate user. They would be several hundred megawatts in size and would use coal as the primary fuel; thus, fuel cell power plants would need to be integrated with a coal gasifier.

In considering the development of a fuel cell power plant for electric utility use, the fuel cell's role must be carefully analyzed, since it will

42 *Arnold P. Fickett*

Fig. 20. Possible fuel cell locations in utility system.

dictate the value to the utility and, hence, the economic viability. For example, the value of a fuel cell power plant fueled by natural gas but without heat recovery is less than the value of the same power plant with heat recovery. Both have lower value than a fuel cell power plant with heat recovery, but fueled by coal (the power plant here would include a coal gasifier). Also, fuel cell power plants located at transmission substations may reduce or defer other utility costs and, therefore, add to their value. They may, for instance, reduce transmission line losses or defer the need for new transmission facilities. As discussed next, all of these factors must be carefully analyzed before establishing the role and specification for a fuel cell system.

Figure 21 depicts a typical electric utility load curve. It is characterized by a relatively low, flat demand during much of the night, with an

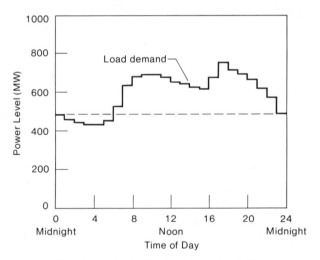

Fig. 21. Typical daily electric load variations.

increasing demand in the early morning and with significant variations during the rest of the day and early evening. Depending on the utility's location and the season, there may be one or two peaks during the day. This curve can be separated into three areas: the "baseload" power level required at least 70% of the time, the incremental power level (called "intermediate duty") above the baseload required 15–70% of the time, and the additional power level required to meet the peaks that will occur 0–15% of the time. Figure 22 shows these requirements as a function of the total peak demand and operating hours per year.

There are two major components to electricity cost: the fuel cost and the (amortization of the) capital cost. For baseload plants, fuel cost is the more significant component, whereas for peakers, capital cost is more important since these units operate for relatively few hours and therefore use relatively little fuel, but charge the (interest on) capital against the few operating hours.

For economical electricity production, a utility carefully plans its generation mix to use low-cost fuels (coal and nuclear) to provide baseload capacity; the high capital cost of the units can be amortized over the long operating hours. Peaking power plants tend to have the lowest capital cost but require expensive liquid and gaseous fuels. Since peakers do not operate for many hours, capital cost is very important, while fuel cost and efficiency are not major concerns. Intermediate duty plants are intermediate in both capital and fuel costs, the latter by virtue of their improved efficiency.

Table VI identifies characteristics of conventional utility power plants. By comparing these with the fuel cell power plant characteris-

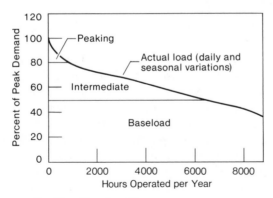

Fig. 22.   Electric utility power requirements.

TABLE VI

Characteristics of Typical Conventional Power Plants[a]

	Baseload generator	
Characteristic	Coal-fired steam	Nuclear
Size (MW)	500–1000	1000
Fuel	Coal	Uranium
Lead time (years)	8	11
Capital cost (1981 $/kW)	1000	1150
Electric efficiency at rated power	36%	32%

	Intermediate-duty generator	
Characteristic	Oil-fired steam	Combined cycle
Size (MW)	500	250
Fuel	Distillate	Distillate
Lead time (years)	7	4
Capital cost (1981 $/kW)	570	430
Electric efficiency at rated power	36%	40%

	Peaking generator
Characteristic	Combustion turbine
Size (MW)	75
Fuel	Distillate/natural gas
Lead time (years)	3
Capital cost (1981 $/kW)	205
Electric efficiency at rated power	30%

[a] Source: EPRI Technical Assessment Guide (1982).

tics (Tables VA and VB), it can be seen that the liquid–gas-consuming fuel cells (Table VA) most closely match the intermediate-duty generator characteristics, and the coal-using fuel cell power plants (Table VB) match the baseload generator characteristics. Figure 23 illustrates the annual cost of operation for the various technologies as a function of operating hours. This very simple curve ignores many of the complexities of a real utility system (such as energy storage, maintenance, reserve requirements, interest during construction, etc.) and, therefore, should be viewed only as an illustration. The most attractive options for each time period are shown as the darkened lines. As can be seen,

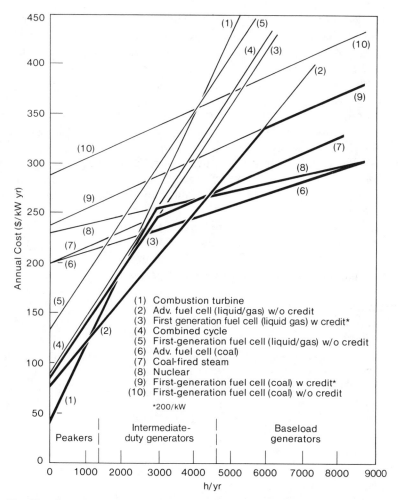

Fig. 23. Annual operating costs versus annual operating hours for various types of generation.

out to about 4000 hours of annual operation, liquid or gaseous fuel technologies offer the minimum annual electricity cost. In these peaking/intermediate-duty regimes, the fuel cell will compete with the combustion turbine and combined cycle plants. Figure 23 also suggests that the first-generation fuel cell is only competitive if assigned credits (see curve 3 versus curve 5) for its unique features, whereas the advanced fuel cell would be competitive even without assigning credits. Beyond 4000 hours, coal or nuclear fuel is the lowest-cost energy source; and fuel cells integrated with coal gasifiers will compete with these technologies. Again, first-generation fuel cells are only competitive if assigned a credit.

The issue of credits or "added values" that accrue from the fuel cell's unique features is controversial. This issue was addressed by the utility planners comprising the System Planning Subcommittee of the Electric Utility Fuel Cell Users Group, who developed a list of benefits that should be considered for dispersed fuel cells; they also set forth low and high values that might be expected from those benefits. Subsequently, 25 utilities reviewed these values and defined the value applicable to their system.

Table VII summarizes the benefits and values as perceived by these utility planners. In addition to those credits shown in Table VII, there are others that would be automatically captured in a utility's conventional planning methodology; that is, whereas Fig. 23 is a screening tool only, a sophisticated computer program would automatically calculate and assign the credit in the course of selecting the lowest-cost generation alternative. For example, the modularity and short lead times reduce the initial capital outlay, which is a financial benefit. Higher reliability and small size reduce the system reserve margin (capacity requirements). Lead time, reliability, efficiency, and size are typical of benefits that would be assigned values by any sophisticated computer program.

Furthermore, if the fuel cell were purchased by a third party and

TABLE VII

Value ($/kW) of Benefits to Be Derived from Dispersed Fuel Cell Power Plants

Benefit	Low value	High value	Average value[a]
Spinning reserve	0	33	8
Load following	0	52	7
VAR control	11	22	12
T&D capacity	0	247	59
T&D energy losses	8	35	15
Cogeneration	0	442	56
Air emission offset	0	217	11

[a] Average of values selected by 25 utility planners.

electricity were sold to the electric utility, then faster depreciation could be used, allowing a tax benefit. This tax benefit could also apply to other forms of generation. However, because the capital outlay for a fuel cell is modest, creative financing options are more manageable and therefore more likely to attract third-party investors.

To summarize the "credit concept" as it relates to fuel cells, the value of $200/kW illustrated in Fig. 23 for both dispersed and central station power plants is reasonable. Credits of at least this magnitude are likely to apply (for different reasons) to both liquid–gas-fueled, intermediate-duty and coal-fueled, baseload fuel cell power plants. However, in a real application, the credits will vary, depending upon the location, the fuel, the existing generation mix, the need for waste heat, the ability to develop third-party financing, and the like. Although advanced fuel cells will be competitive without credits, the first-generation fuel cell will probably only be used in situations in which credits or "added value" can be derived.

C. Fuel Scenarios

The widespread acceptance of fuel cell power plants by the electric utility industry requires the assurance of a fuel supply that is both economical and logistically feasible, that is, a supply that utilities can order in large quantities from conventional fuel companies. Although fuel cell power plants will be able to utilize any liquid or gaseous fuel and are no different from many other utility power plants in this regard, fuel cell developers have been saddled with the burden of responding to the anxieties of the potential users and the funding agencies concerning fuel cost and availability. These anxieties have resulted from a combination of many factors, including the following.

1. The dramatic increase in the price of petroleum that started with the formation of OPEC in the early 1970s and continued throughout that decade.
2. The declaration by the Carter administration in 1976–1977 that the electric utility industry would be constrained in the use of petroleum and natural gas. This requirement was subsequently implemented under the Industrial Fuel Use Act of 1978. Although the act specifically exempted fuel cells, the utility industry was not completely convinced that this exemption would be permanent.

3. The initiation of many synthetic fuel projects throughout the 1970s in response to the perceived petroleum and natural gas shortages. This emphasis on synthetic fuels led many to believe that such fuels should be readily available by the mid-1980s.
4. An escalation of power plant emission regulations that restricted the siting of power plants, especially coal-fired power plants but also plants fueled by petroleum and natural gas that emitted SO_x and NO_x in unacceptable quantities.

In effect, many electric utilities found themselves in a *Catch 22* situation regarding fuel use in power plants. Even the use of coal, the most acceptable fuel under the Fuel Use Act, resulted in siting and environmental restrictions that greatly increased the cost of power plants and added years to the permit and licensing cycle. Thus, it is not surprising that in considering an alternative generation technology, the industry would require an answer to the fuel question.

Of course, the fuel situation has again changed. Natural gas and petroleum are, temporarily at least, in good supply. Consequently, the Fuel Use Act is not rigidly enforced. Furthermore, many of the synthetic fuel projects initiated in the 1970s are being terminated for lack of economic viability.

Given the vast changes in the fuel scenarios that have transpired over the past 20 years, we must conclude that developing a technology to meet any single future scenario would be foolhardy. Consider, for example, the following possible scenarios, any of which has some popular support based on a rational argument.

1. A Middle East war will cause new petroleum shortages and price increases.
2. There exist large supplies of natural gas trapped during the earth's formation that will provide virtually limitless quantities at depths that are just being reached by advanced drilling rigs.
3. OPEC initiates a price war, causing dramatic petroleum price reductions.
4. Concerns regarding the greenhouse effect result in further environmental constraints regarding fossil fuel use and CO_2 emissions.
5. The world economy rapidly improves in the middle 1980s, placing new demands on fuel and energy sources.

These are but five of many scenarios that might be offered. All have a reasonable probability of occurrence but would have a dramatically different impact on the electric utility industry and the cost and availability of fuel.

In an attempt to delineate the perceptions of the utilities regarding fuel availability, a survey of more than 30 utilities was conducted by the Fuels Subcommittee of the Electric Utility Fuel Cell Users Group (1980). Table VIII summarizes the results of this survey. As can be seen, methane is perceived to be generally available through 1990 as natural gas and LNG with possible availability as SNG occurring after 1990. Petroleum liquids are seen to be available into the 1990s. Methanol and medium-British thermal unit coal gas are perceived to have some probability of becoming available in the 1990s; however, there is no consensus regarding the availability of many of the synthetic fuels.

As a result of this survey, the electric utility fuel cell community (developers, utilities represented by the Fuel Cell Users Group, and funding organizations) has adopted the following fuel strategy.

1. Methane will be considered as the baseline fuel: It is available in the near term as natural gas; it is an opportunistic fuel for special situations, and it can be obtained from solid waste, secondary sewage, and biomass; in the longer term, SNG could be obtained from coal conversion.
2. The fuel cell power plant should be capable of low-cost modification to use any other clean liquid or gaseous fuel, such as naphtha, methanol, or coal gas (ranging from low to high British thermal units).
3. Fuel cell power plants integrated with coal gasifiers will be considered as alternatives to liquid–gas-fueled power plants: these plants are best suited to baseload, central station applications; they offer the ultimate in an environmentally acceptable, coal-fueled power plant.

The consequence of this strategy on the first-generation fuel cell power plant can be seen in Table IX, which depicts the impact of modifying a methane-fueled power plant for alternative fuels. As can be seen, the impact is relatively minor, confirming the generally held belief that fuel flexibility is truly one of the fuel cell's attributes. It is believed that similar fuel flexibility would be a part of a molten carbonate or a solid oxide power plant.

TABLE VIII

Perceived Fuel Availability: Number of Utilities Responding by Rating Number (Fuel Cell Users Group, 1980)

Fuel	Rating[a]	1981–1985	1986–1990	1991–2000	+2000
			Time frame		
Methane as natural gas, LNG	1	23	18	3	0
	2	5	8	19	17
	3	5	7	11	16
Methane as SNG	1	0	0	0	4
	2	2	6	27	29
	3	31	27	6	0
Naphtha, kerosene, or jet fuel	1	23	20	13	0
	2	6	8	13	23
	3	4	5	7	10
#2 fuel oil	1	29	22	12	0
	2	3	9	18	22
	3	1	2	3	11
#4 and #6 fuel oils	1	28	26	22	0
	2	3	5	8	27
	3	2	2	3	6
Light coal—liquids	1	0	0	0	2
	2	0	0	11	18
	3	33	33	22	13
Medium-Btu gas from coal	1	0	1	3	10
	2	0	16	27	22
	3	33	16	3	1
Low-Btu gas from coal	1	0	0	2	7
	2	0	20	10	9
	3	33	31	21	17
Methanol	1	0	1	1	5
	2	2	17	20	28
	3	31	15	3	0
Ethanol	1	0	3	3	5
	2	4	6	13	14
	3	29	24	17	14
Liquified petroleum gases (LPG)	1	0	0	0	0
	2	6	7	6	5
	3	27	26	27	28
Hydrogen	1	0	0	0	2
	2	1	4	6	9
	3	32	29	27	22
Heavy coal liquids	1	0	0	0	0
	2	0	2	14	28
	3	33	31	19	5

[a] Rating: 1, high probability that fuel will be available and used by respondent; 2, possible availability; 3, high probability that fuel will neither be available nor used by respondent.

TABLE IX

Impact of Retrofitting a Fuel Cell Power Plant for Use with Alternative Fuels[a]

Fuel	Efficiency (%)	Capital cost increase (%)
Methane (baseline)	40.9	0
Naphtha	41.1	0
Medium-Btu Gas		
With methane	40.8	+0.4
Without methane	40.9	+4.3
Methanol	41.6	+1.0
Hydrogen	42.1	+1.1

[a] Source: TVA (Jackson, 1981).

Table X presents a projection (Holtberg *et al.* 1982) of the prices of the primary fuels that might be used in electric utility fuel cell power plants.

The synthetic fuels are expected to move into the fuel supply system at or near the price of the equivalent natural fuel. For instance, methanol would be expected to compete with petroleum fuels only if it were available at an equivalent price. Likewise, coal gases will need to be priced the same as, or lower than, natural gas before they will be competitive.

TABLE X

Projection of Prices of Primary Fuels for Electric Utilities

Fuel	Fuel price (1981 $/$10^6$ Btu)		
	1981	1990	2000
Natural gas	2.80	5.19	7.11
Naphtha	7.68	8.17	9.94
Distillate	7.21	7.64	9.31
Coal	1.51	2.08	2.56

D. Integration within Fuel Cell Power Plants

Figure 2 depicted a simplified representation of a fuel cell power plant made up of the fuel processor, fuel cell power section (cell stacks), and power conditioner. In actuality, the power plant is much more complex than Fig. 2 would suggest, and great care is taken to optimize thermal integration within the power plant in order to maximize its overall efficiency. This overall efficiency E_T is determined as

$$E_T = E_{FP}E_{FC}E_{PC},$$

where E_{FP} is fuel processor efficiency, which equals the lower heating value of the fuel exiting the fuel processor divided by the higher heating value of the fuel entering the fuel processor; E_{FC} is fuel cell efficiency, which equals the dc power produced by fuel cell divided by the lower heating value of the fuel exiting the fuel processor; E_{PC} is power conditioner efficiency, which equals the ac power produced by the power conditioner divided by the dc power produced by the fuel cell; and E_T equals the ac power produced by the power conditioner divided by the higher heating value of the fuel entering the fuel processor.

Fuel cell efficiency can be further broken down into its components:

1. a voltage efficiency E_V that is the ratio of the actual cell voltage over that which is theoretically attainable (see Figs. 3 and 6),
2. a fuel utilization efficiency E_F that is the ratio of the fuel electrochemically used by the fuel cell over that available to the fuel cell, and
3. a thermodynamic efficiency term E_{Th} that accounts for the fact that not all of the fuel's enthalpy can be converted to electrical energy. [In fact, the thermodynamic limit is the ratio of Gibbs free energy ΔG to enthalpy ΔH. Thus, $E_{Th} = \Delta G/\Delta H$ for the equation $H_2 + \frac{1}{2}O_2 \rightarrow H_2O_{(g)}$].

Typically, the efficiencies for a fuel cell power plant would break down as

$$E_{FP} \simeq 88\%,$$

$$E_{FC} \simeq E_V E_F E_{Th} \simeq (0.72/1.2) \times 0.85 \times 0.93 \times 100 \simeq 47\%,$$

$$E_{PC} \simeq 95\%,$$

$$E_T \simeq 0.88 \times 0.47 \times 0.95 \times 100 \simeq 40\%.$$

`To achieve this and higher overall efficiencies, thermal integration between the subsystems is necessary.

In cogeneration applications in which the waste heat is used to produce thermal energy, the thermal energy must be accounted for in the overall efficiency calculation. As a rough approximation, about as much thermal energy as electrical energy can be derived from a fuel cell power plant. The quality of the thermal energy will vary, depending on the fuel cell type. The PAFC power plant being developed by UTC (Handley and Cohen, 1981) would produce 11-MW electric energy and 42×10^6 Btu/h of 200–250°F hot water.

Figure 24 depicts a fuel cell power plant system schematic that is still quite simplified but begins to show the integration within the power plant as well as the interface for power plant cooling or a cogeneration option.

At this level of integration the simple approach to bookkeeping of subsystem efficiency breaks down as waste heat from one subsystem is transferred and used in another. Thus, in reality a very rigorous heat and mass balance must be developed and optimized at the component level; this must be continuously reiterated to ensure that the overall power plant is optimized.

1. Phosphoric Acid Fuel Cell Systems

Figure 24 depicts the PAFC power plant as it is presently envisaged. It is representative of the phosphoric acid power plant whose characteristics were shown in Table VA. In this power plant, fuel entering the plant is pumped to system pressure, vaporized, and mixed with a hydrogen-rich recycle stream. This mixture passes through a hydrodesulfurizer, which converts the fuel's sulfur compounds into hydrogen sulfide (H_2S). The H_2S is removed by adsorption on a zinc oxide bed. The desulfurized fuel is then combined with steam and enters the reformer, where the mixture is catalytically converted into a hydrogen-rich gas. The H_2 content of the product gas is further increased by two stages of shift conversion ($H_2O + CO \rightarrow CO_2 + H_2$). The gas, which is then cooled, passes into the power section (cell stacks), where hydrogen and oxygen from the process gas and air streams are electrochemically combined, producing dc electricity and by-product water. Both the air and fuel streams entering the fuel cell are pressurized. The depleted fuel gas stream leaves the power section and passes through a water-recovery condenser. After leaving the condenser, it enters the reformer burner, where it is combusted to provide thermal energy for

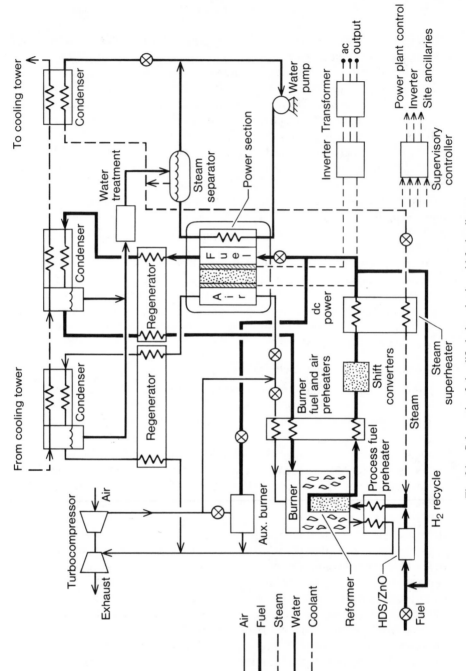

Fig. 24. Schematic of simplified phosphoric acid fuel cell system.

the steam-reforming reaction. Process air leaving the power section contains by-product water, which is recovered by a condenser. The dried, pressurized air is mixed with the hot, high-pressure reformer burner exhaust. This combined stream is then expanded through a power recovery turbine which, in turn, drives the process air compressor.

Water recovered in the condensers is recycled to the thermal management subsystem. The primary function is to control the power section temperature by circulating water (steam) through the cell stacks. Waste heat generated in the process of producing power is removed by evaporating a portion of the circulating water. Steam is separated for use in the reformer. The remaining water is collected, purified, and circulated through the cell stacks. Power (dc) from the fuel cells is converted to utility-quality power (ac) by a self-commutated inverter and conventional transformer.

Figure 25 depicts an artist's rendering of the PAFC power plant described in Table VA.

A PAFC power plant integrated with a coal gasifier (instead of a steam reformer) is shown schematically in Fig. 26 (Cronin *et al.*, 1982). This alternative combines small commercial (Wellman-Galusha) coal gasifier technology with the PAFC. Coal (lignite) is sized and then gravity fed to the 10-ft-diameter fixed bed gasifiers. Air is supplied by a blower, and steam is obtained from the fuel cell coolant system. The coal, air, and steam react in the gasifier to produce carbon monoxide and hydrogen in addition to methane and a variety of tars and oils, as well as sulfur- and nitrogen-bearing components. The gas leaves the upper section of the gasifier at approximately 160°C and atmospheric pressure. The gas passes through cyclones to remove particles and then through the gas cleaning (to remove tars and oils) and cooling sections before being compressed to the power section operating pressure. Compression power is provided by the turboexpansion of the high-temperature flue gas from catalytic combustion of the fuel cell anode vent products plus the cathode vent. The pressurized gas is then cooled and passed through the sulfur removal system, which extracts the bulk of the sulfur in the 99%-pure, molten form. The gas is then passed over zinc oxide and stripped of any remaining sulfur compounds before entering shift reactors to convert the bulk of the carbon monoxide to hydrogen and carbon dioxide. The resulting hydrogen-rich gas is ready for use in the power section.

The coal gasification and subsequent gas cleanup subsystems are much more costly than their counterparts in the natural gas- or petro-

Fig. 25. Artist's rendering of phosphoric acid power plant layout. (Courtesy of United Technologies.)

leum-reforming process, which accounts for the very large difference in capital cost shown in tables VA and VB. Due to the economies of scale of the sulfur-removal processes, power plants smaller than 45 MW are probably not viable. Furthermore, assessments of the PAFC integrated with the coal gasifier are still at an early (conceptual) stage, and more effort is required to verify that the projections (Table VB) are realistic. The hope is that these small coal-fueled power plants would offer a low-cost, dispersed baseload generation alternative.

2. Molten Carbonate Fuel Cell Systems

The MCFC power plant based on a natural gas–petroleum-reforming process would, in principle, resemble Fig. 24 with one exception: the cathode of the molten carbonate fuel cell requires a source of carbon dioxide, since the half-cell reaction is $CO_2 + \frac{1}{2}O_2 + 2e \rightarrow CO_3$. To accomplish this, the anode vent is combusted to convert all unused fuel to carbon dioxide and water; the water is condensed, and the carbon

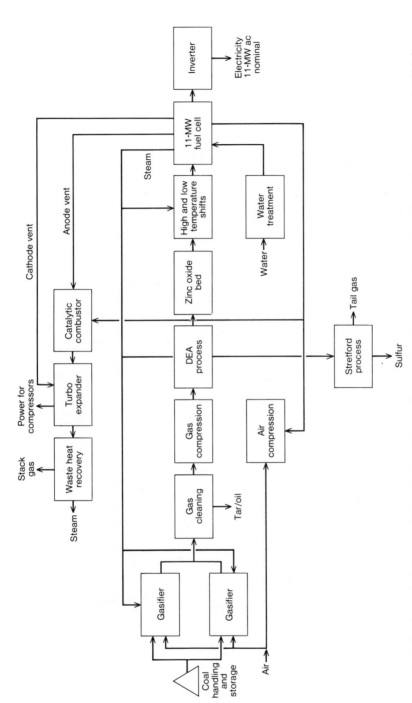

Fig. 26. Conceptual diagram of phosphoric acid fuel cell power plant integrated with a coal gasifier. [From Cronin *et al.* (1982).]

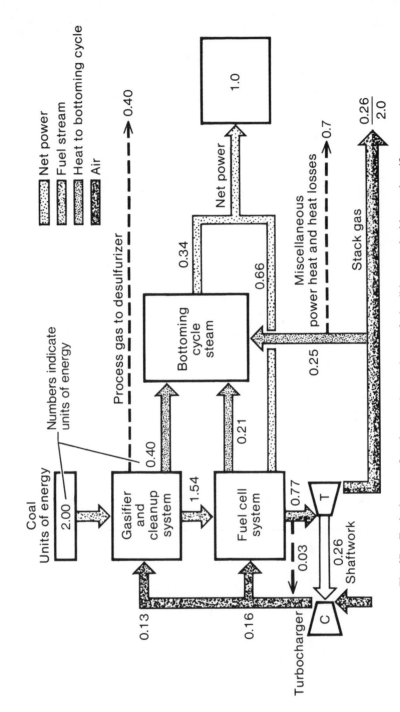

Fig. 27. Typical energy flow for a molten carbonate fuel cell integrated with a coal gasifier.

dioxide is injected into the air feed. Compared to the PAFC, little thought has been given to MCFC power plants based upon natural gas or liquid fuels for two reasons: First, power plants that use gaseous or liquid fuels will likely be deployed for peaking or intermediate duty, which implies that they will start up and shut down at least once per day. This mode of operation is better suited to the lower-temperature PAFC than to the higher-temperature MCFC system. Second, the MCFC, because of its higher operating temperature, would appear to integrate very well with large central station coal gasifiers. In fact, this combination—advanced coal gasifiers and MCFCs—is projected to have the highest efficiency of any coal-fueled system. Figure 27 illustrates a typical MCFC–coal gasifier concept. As can be seen, this concept introduces two units of energy as coal (higher heating value) and delivers 0.66 units of energy from the fuel cell and 0.34 units from the steam bottoming cycle for an overall electrical efficiency of 50%.

Figure 28 shows the schematic of a 675-MW MCFC power plant that was analyzed by General Electric (Bonds and Dawes, 1979). This particular system uses an oxygen-blown Texaco gasifier fed with a coal–water slurry. The coal gas is produced at 1370°C. The gasifier effluent passes through a high-temperature steam generator, dropping the temperature to 650°C, and then through a regenerative heat exchanger train that cools the gas to 38°C, condensing the water. The cool gas proceeds through an NH_3 scrubber, COS converter, and sulfur cleanup. The clean gas leaves the cleanup system at 25°C and is reheated (by the regenerative heat exchanger train) to 620°C before being expanded through the turbine to the fuel cell pressure. After reheating, the gas enters the cell stacks, which operate at a fuel utilization rate of 0.85; that is, 85% of the H_2 and CO are converted to electricity, with the excess exhausted from the stack. The fuel exhaust is catalytically burned and mixed with reaction air (compressed by energy from the clean gas "letdown" turbine) to provide a proper CO_2/O_2 ratio for the fuel cell cathode. This cathode inlet gas is heated to 540°C and then supplied to the fuel cell stacks at an oxygen utilization of 25%. The cathode exhaust gas powers a simple gas turbine with a pressure ratio of six and discharge temperature of 390°C that produces about 75 MW(e). In addition, a steam generator (operation from the various high-pressure steam flows) produces about 150 MW(e). A possible power plant configuration would comprise the following:

1. 600 MCFC stacks (having 500, 1-m² cells each) for a total 450-MW output,

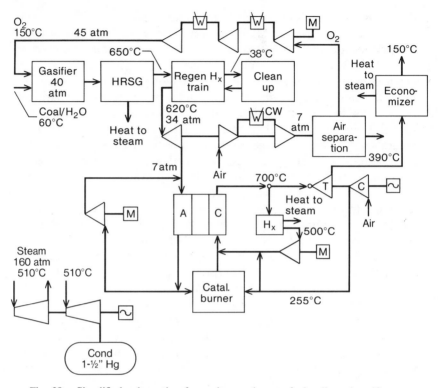

Fig. 28. Simplified schematic of a molten carbonate fuel cell–coal gasifier power plant. [From Bonds and Dawes (1979).]

2. 15 coal gasifiers capable of handling 3×10^8 Btu/h each, 10 heat-recovery steam generators,
3. 5 15-MW gas turbines, and
4. 1 150-MW steam turbine.

Other alternatives have been considered (Bonds and Dawes, 1981), including other gasifiers, cleanup systems, and bottoming cycle arrangements. While it is likely that still other configurations will be developed that are improvements over the preceding, these improvements will probably not be dramatic because the overall power plant efficiency and resulting bus bar cost are relatively insensitive to the many variables.

A MCFC power plant concept that has received recent attention involves alternatives to the conventional reforming of natural gas in a

MCFC power plant. Since the MCFC operates at a temperature at which natural gas reforms readily, it is possible to consider reforming the natural gas within the anode compartment of the MCFC. This results in certain advantages.

1. The fuel cell waste heat can be used to drive the reforming process directly.
2. The complexity of external fuel-processing equipment is eliminated.
3. Much smaller, modular MCFC power plants can be envisaged.

Various "internal reforming" or "sensible heat reforming" configurations have been conceived (Krumpelt *et al.,* 1982) that project overall power plant efficiencies of 60%. This concept warrants careful watching for progress over the next 2 to 3 years. Even if successful, however, this approach will be limited to use with fuels containing a single carbon atom, i.e., methane (natural gas) and methanol. Fuels containing more than a single carbon do not reform cleanly, and the by-products would eventually contaminate the cell.

3. Solid Oxide Fuel Cell Systems

Very little attention has been devoted to SOFC power plants. Because they operate at very high temperatures, they, like MCFC, are best suited to baseload operation. The cathode requires only an oxygen feed; thus, in principle, the system can be simpler than that of the MCFC, which needs to transfer carbon dioxide to the air inlet. Otherwise, the power plant can be expected to be similar to MCFC power plants. This is even true of the internal reforming concept, which can apply to the SOFC as well as to the MCFC.

E. State of Hardware Development

1. Phosphoric Acid Fuel Cells

There are two major PAFC development efforts going on in the United States that are focused on the electric utility application: that of United Technologies Corporation (UTC) and another involving a Westinghouse/Energy Research Corporation (W/ERC) team. The PAFC technologies being pursued by these developers are very similar, with the only significant difference being the means of stack cooling. United Technologies Corporation removes heat by circulating a two-phase

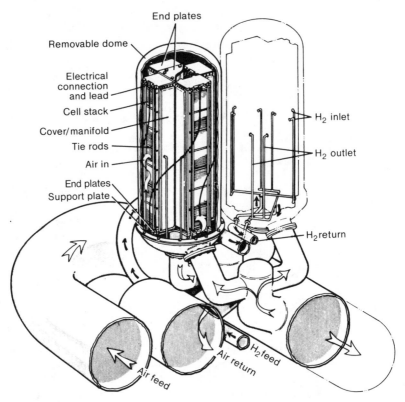

Fig. 29. Artist's rendering of an air-cooled phosphoric acid fuel cell module. (Courtesy of Westinghouse.)

water-steam mixture through coolant plates located between blocks of four to eight cells in the stack. Westinghouse/Energy Research Corporation uses a patented distributed gas (DIGAS) concept that allows a portion of the reaction air stream to pass through coolant plates similarly located between blocks of cells. The UTC approach offers slightly better efficiency and smaller air inlet ducts. The W/ERC concept offers a simpler design with less internal stack plumbing. Figure 29 is an artist's concept of a DIGAS module.

The UTC effort has resulted in the following hardware accomplishments of note:

1. the fabrication and operation in 1976–1977 of a 1-MW pilot plant for about 1500 hours,

2. the fabrication and installation of a 4.5-MW (ac) demonstration power plant in New York City in the Consolidated Edison system, and
3. the fabrication and installation of a similar 4.5-MW (ac) demonstrator in Tokyo, Japan, in the Tokyo Electric Power system.

The New York City demonstrator is depicted in Fig. 30. The principal difference between the two 4.5-MW units is that the Tokyo unit was fabricated some 2 to 3 years after the New York unit and has a more advanced cell stack that offers 40,000 hours of endurance. The technology available in 1976 when the New York unit was commissioned was limited to a few thousand hours. Figures 31 and 32 illustrate the PAFC progress that has occurred since 1976 in terms of both life and performance. The Tokyo unit contains a stack representative of the better-performing and longer-lived 1978–1979 technology.

The preceding activities were implemented to confirm the power plant concept, to demonstrate sitability, to verify operational characteristics of components and system interfaces, and to obtain utility experience with operation and maintenance. In addition to these activi-

Fig. 30.   Artist's rendering of the 4.5-MW (ac) demonstrator in New York City.

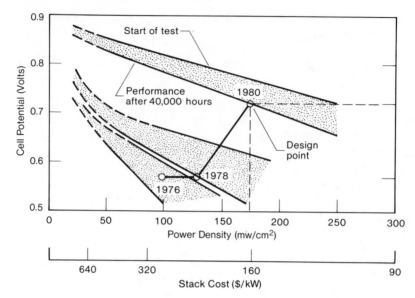

Fig. 31.   Progress in phosphoric acid fuel cell technology.

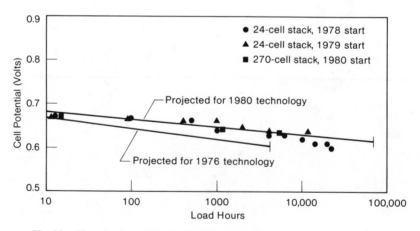

Fig. 32.   Phosphoric acid fuel cell stack performance versus operating time.

ties, there is a parallel effort to result in the design, development, and component-level verification of a commercial prototype power plant. This prototype is characterized in Table VA (first generation) and will contain many improvements over the 4.5-MW units:

1. higher-temperature, higher-pressure operation,
2. improved efficiency—41% rather than the 4.5-MW demonstrator's 37% efficiency,
3. fewer parts—a reduction of about 35% in the parts count,
4. more "off-the-shelf" rather than customized components, and
5. improved maintainability/accessibility.

The W/ERC effort is relatively new and has progressed from the test of small fractional kilowatt stacks to 2- and 10-kW stacks. Paralleling its development of a technology base, the W/ERC team has also embarked upon the design, development, and verification of a commercial prototype. The present W/ERC "centerline" design is for a 7.0-MW (ac) power plant with 40% electrical efficiency. Small stacks have operated for up to 20,000 hours with electrolyte (acid) replenishment. The W/ERC plan calls for the test of a 100-kW stack in 1984 and a 400-kW module in 1985.

Figure 33 summarizes the timeline for the projected UTC and W/ERC activities.

In addition to the U.S. activities, there is now a rapidly accelerating PAFC development program in Japan. Five Japanese companies—Mitsubishi, Fuji, Toshiba, Hitachi, and Sanyo—are involved in PAFC research and development. The first four mentioned are part of the Moonlight Project effort to install and test a 1-MW power plant in about 1986. Although this may appear to be late compared to the U.S. activities noted in Fig. 33, it must be realized that the Japanese effort is very new. Thus, their progress to date, which includes the fabrication and test of a 30-kW power plant by Fuji, is impressive.

2. Molten Carbonate Fuel Cells

There are three major MCFC developers in the U.S. who are focusing on electric utility power plants: United Technologies Corporation (UTC), General Electric (GE), and Energy Research Corporation (ERC). The fundamental cell stack technology is still in a state of flux. The largest unit that has been tested is a stack of 20 1-ft^2 cells similar to that shown in Fig. 34. This stack was operated by UTC for about 900 hours at ambient pressure and a subsequent 1100 hours at pressures up

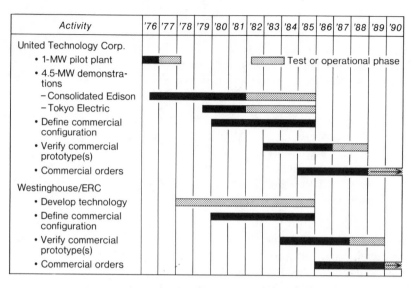

Fig. 33. Schedule of phosphoric acid fuel cell activities.

Fig. 34. Twenty-cell molten carbonate fuel cell stack. (Courtesy of United Technologies.)

to 4 atm. This stack was followed by a series of 8- to 10-cell stacks to evaluate improved cell and stack components. The most successful of these tests is depicted in Fig. 35. This result was intended to lead to the release of a 100-cell stack that would operate in a "breadboard" system, including reformer and other key system ancillaries. The nominal output of this system is 10 kW (dc). However, testing of additional 10-cell stacks revealed new problems:

1. Only a small portion of the carbonate electrolyte that is lost from the cells during operation can be accounted for. This issue is important to understanding and ensuring long-term endurance.
2. The nickel oxide cathode was found to dissolve at pressurized operation. This problem must be resolved for long-term, pressurized operation.

As a consequence, the fabrication of the 10-kW system was postponed, awaiting resolution of these issues (King *et al.,* 1982).

Although neither GE nor ERC have yet tested multicell stacks, both have initiated stack design activities and can be expected to move soon into the stack development phase.

There are now significant MCFC development efforts in both Japan and Italy.

3. Solid Oxide Fuel Cell

There is presently only a single U.S. SOFC developer—Westinghouse. As discussed earlier, a 10-cell stack had been operated for a

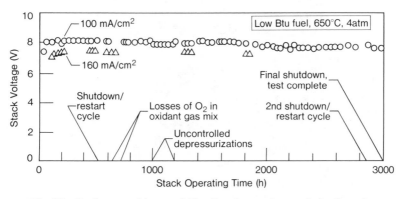

Fig. 35. Performance history of 10-cell molten carbonate fuel cell stack.

total of 5000 hours. Since that test, however, Westinghouse has developed a new stack concept (Isenberg, 1982). Whereas the previous concept placed many cells connected in series on a single tube, the new concept (Fig. 36) employs only one cell per tube but then interconnects the individual tubes in series/parallel arrays to achieve the desired output. There are no reported test results involving this stack concept at this time.

As in the case of the MCFC, there are no notable SOFC activities taking place outside of the United States.

F. Prognosis

The prospects for electric utility fuel cell power plants would seem to be good. Fuel cells offer many advantages ranging from the benefits derived from dispersed fuel cell power plants, as presented in Table VII, to the environmental and efficiency benefits that could be derived from future baseloaded, central station power plants, as summarized in Table VB.

In addition, the fuel question that plagued the PAFC program for

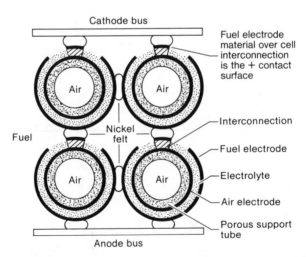

Fig. 36.   Schematic of solid oxide fuel cell bundle. [From Isenberg (1982).]

almost a decade seems to have been resolved by (1) the ability of present PAFC power plant designs to use virtually any clean liquid or gas and (2) the current perception that clean liquid and gaseous fuels will be available to electric utilities for the foreseeable future.

An indication of the electric utility interest in fuel cells is the formation of the Fuel Cell Users Group by more than 50 utilities. The Fuel Cell Users Group's primary purpose is to expedite the commercial introduction of PAFC power plants. Similarly, both UTC and GE have formed advisory committees for their MCFC programs. These committees involve another dozen or so utility representatives.

An assessment of the market for dispersed PAFC power plants was carried out under the direction of the Fuel Cell Users Group. This effort utilized a utility generation expansion model and methodology to determine the market potential for fuel cells in 25 electric utilities, who produce over 15% of the electric energy in the United States. Figure 37 portrays the results of that effort. This figure suggests that, with a credit of $100–200/kW and an installed cost of $600/kW, there could be a market of about 15,000 MW for the 25 utilities and about 80,000 MW

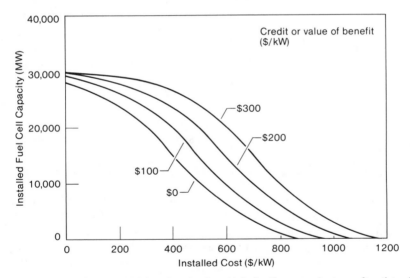

Fig. 37. Market potential for phosphoric acid fuel cell power plant as a function of cost.

for the United States (assuming that the 25 are typical). Even at \$800/kW there would be a significant opportunity. These results indicate that if the developers can meet their goals, then first-generation, dispersed PAFC power plants will move into the market. These will be followed by advanced PAFC power plants as well as by MCFC and SOFC power plants in due time. The real issue, then, is whether the developers can meet their goals, which are no longer technical but are now actually cost goals. There seems to be little doubt as to whether the PAFC power plant can be made to meet performance and endurance targets. There is real concern, however, as to its ability to meet the cost targets. The cost problem arises, in part, because fuel cells are a factory-produced technology. By its very nature, factory production is meaningless unless large numbers of units are involved. The \$600/kW fuel cell capital cost assumes that 500 MW/year will be produced. To reach that production rate, many utilities will need to place orders. However, utilities are not likely to place many orders until they can observe a number of commercial units in operation. Thus, a *Catch 22* situation presents itself: In limited production, fuel cells are virtually custom made and cost three to four times what they could cost in production quantities. Yet, utilities are unlikely to order large quantities without seeing several commercial units in operation, and manufacturers are unlikely to risk venture capital without some indication of utility orders.

Before fuel cells can become commercial, this *Catch 22* cycle will need to be interrupted through some combination of the following factors:

1. manufacturers' willingness to risk capital and/or forward pricing,
2. users' (utilities') willingness to assume risk and/or premium cost for early units, and
3. government/EPRI willingness to support (subsidize) early units.

If one or more of these situations occur and electric demand continues to grow at a modest rate, then fuel cells should begin to find their way onto utility systems late in the decade. Once fuel cells begin to penetrate the marketplace, improved versions will follow through an evolution of the first-generation PFC power plants, as well as through the introduction of new advanced power plants as they become available in the middle to late 1990s.

REFERENCES

Abens, S. G., Hofbauer, J. A., and Marchetto, P. G. (1981). 3 and 5 kW Methanol power plant program. *Abstr. Natl. Fuel Cell Semin.,* Courtesy Associates, Washington, D.C.

Adlhart, O. J. (1978). An assessment of the air breathing, hydrogen fueled SPE cell. *Proc. 28th Power Sources Symp.,* Redbank, New Jersey.

Bacon, F. T. (1954). Fuel cells. *BEAMA J.* **61.**

Barthelemy, R. (1981). Defense applications of fuel cells. *Abstr. Natl. Fuel Semin.,* Courtesy Associates, Washington, D.C.

Benjamin, T. G., Camara, E. H., and Marianowski, L. G. (1980). "Handbook of Fuel Cell Performance." Prepared for the U.S. Department of Energy, Contract No. EC77C03-1545, Institute of Gas Technology.

Bett, J. S., Bushnell, C. L., and Buswell, R. F. (1980). "Advanced Technology Fuel Cell Program." Electric Power Research Institute Project 114-2, Annual Report No. EM1328. Power Systems Division of United Technologies Corporation.

Bonds, T. L., and Dawes, M. H. (1979). "Fuel Cell Power Plant Integrated Systems Evaluation." Electric Power Research Institute Project 1085-1, Interim Report No. EM1097. General Electric Company.

Bonds, T. L., and Dawes, M. H. (1981). "Fuel Cell Power Plant Integrated Systems Evaluation." Electric Power Research Institute Project 1085-1, Final Report No. EM1670. General Electric Company.

Cronin, P. G., Murphy, A. J., Newton, R. J., and Wagner, E. S. (1982). "Assessment of a Coal Gasification Fuel Cell System for Utility Application." Electric Power Research Institute Project 1041-8, Final Report No. EM2387. Kinetics Technology International.

Electric Power Research Institute (1982). "Technical Assessment Guide."

Fickett, A. P. (1977). "Fuel Cell Electrocatalysts—Where Have We Failed?" Paper presented at Spring Meeting Electrochem. Soc., Philadelphia, Pennsylvania.

Fuel Cell Users Group (1980). "Report to the Management Committee of the Electric Utility Fuel Cell Users Group." Prepared by Fuels Subcommittee.

George, M., and Scozzofava, J. (1978). Reversible metal hydride–air fuel cell. ECOM Report 77-2644-F. Ft. Monmouth, New Jersey.

Grove, W. R. (1839). On voltaic series in combinations of gases by platinum. *Philos. Mag.* **14,** 127–130.

Grove, W. R. (1842). On a gaseous voltaic battery. *Philos. Mag.* **21,** 287–293.

Grove, W. R. (1874). "The Correlation of Physical Forces," 6th ed. Longmans, Green, New York.

Grubb, W. T. (1957). Ion exchange batteries. *Proc. 11th Annu. Battery Res. Dev. Conf.,* pp. 5–8. Atlantic City, New Jersey.

Handley, L. M., and Cohen, R. (1981). "Specification for Dispersed Fuel Cell Generator." Electric Power Research Institute Project 1777-1, Interim Report No. EM2123. Power Systems Division of United Technologies Corporation.

Holtberg, P. D., Woods, T. J., Hill, R. H., and Rasmussen, J. J. (1982). "1982 GRI Baseline Projection of U.S. Energy Supply and Demand, 1981–2000." Gas Research Insights.

Houghtby, W. E., King, J. M., and Thompson, R. A. (1978). ''Advanced Technology Fuel Cell Program.'' Electric Power Research Institute Project 114-2, Annual Report No. EM956. Power Systems Division of United Technologies Corporation.

Isenberg, A. O. (1980). Processing and performance of high temperature solid oxide fuel cells. *Abstr. Natl. Fuel Cell Semin.* Courtesy Associates, Washington, D.C.

Isenberg, A. O. (1982). Recent advancements in solid electrolyte fuel cell technology. *Abstr. Natl. Fuel Cell Semin.* Courtesy Associates, Washington, D.C.

Jackson, S. B. (1981). Performance and cost impacts of using coal-derived fuel in phosphoric acid fuel cell power plants. *Abstr. Natl. Fuel Cell Semin.* Courtesy Associates, Washington, D.C.

King, J. M., Reiser, C. A., and Schroll, C. R. (1982). ''Molten Carbonate Fuel Cell Systems Verification and Scale-Up.'' Electric Power Research Institute Project 1273-1, Interim Report No. EM2502. United States Technologies Corporation.

Krumpelt, M., Ackerman, J., Herceg, J., Zwick, S., Slack, C., and Lwin, Y. (1982). Gas systems. *Abstr. Natl. Fuel Cell Semin.,* Courtesy Associates, Washington, D.C.

Kudo, T., and Obayashi, H. (1976). Ion–electron mixed conduction in the fluorite type $Ce_{1-x}Gd_xO_{2-x/2}$. *J. Electrochem. Soc.* **123**, 415.

Kunz, J. R., and Gruver, G. A. (1975). The catalytic activity of platinum supported on carbon for electrochemical oxygen reduction in phosphoric acid. *J. Electrochem. Soc.* **122**, 1279.

Lebhafsky, H. A., and Cairns, E. J. (1968). ''Fuel Cells and Fuel Cell Batteries,'' pp. 18–47. Wiley, New York.

Lindstrom, O. (1964). Fuel cells. *ASEA J.* **37**, 3.

Lorton, G. P. (1980). ''Sulfur Removal Processes for Advanced Fuel Cell Systems.'' Electric Power Research Institute Project 1041-5, Final Report EM1333. C. F. Braun.

McBryar, H. (1979). ''NASA fuel cell program plan.'' *Abstr. Natl. Fuel Cell Semin.* Courtesy Associates, Washington, D.C.

McCormick, J. B., Huff, J., Srinivasan, S., and Bobbett, R. (1979). ''Application Scenario for Fuel Cells in Transportation.'' Report LA 7634-MS, Los Alamos National Laboratory.

Mientek, A. P. (1982). ''On Site Fuel Cell Power Plant Technology Development Program.'' Prepared for GRI by United Technology Corporation, South Windsor, Connecticut.

Rohr, F. S. (1977). High temperature solid oxide fuel cells—Present state and problems of development. *Ext. Abstr. Workshop High Temp. Solid Oxide Fuel Cells.* Brookhaven National Laboratory, New York.

Ross, P. N., Jr., and Benjamin, T. G. (1977). Thermal efficiency of solid electrolyte fuel cells with mixed conduction. *J. Power Sources* (January), 311–321.

Ross, P. N., Jr., (1980). ''Oxygen Reduction on Supported Pt Alloys, and Intermetallic Compounds in Phosphoric Acid.'' Electric Power Research Institute Project 1200-5, Final Report No. EM-1553. Lawrence Berkeley Laboratories, Berkeley, California.

Stonehart, P., and Baris, J. (1980). Preparation and evaluation of advanced electrocatalysts for phosphoric acid fuel cells. *First Q. Rep. NASA CR 159843.*

Stonehart, P., and MacDonald, J. P. (1981). ''Stability of Acid Fuel Cell Cathode Materials." Electric Power Research Institute Project 1200-2, Interim Report No. EM1664. Stonehart Associates.

Strasser, J. (1979). Development of a 7 kW H_2/O_2 fuel cell assembly with circulatory electrolyte in a compact modular design. *Proc. Electrochem. Soc.* Boston, Massachusetts.

Solar Ponds

N. D. Kaushika

Centre for Energy Studies
Indian Institute of Technology
New Delhi, India

I. INTRODUCTION

The solar pond is a unique development in renewable energy resource technology. It utilizes a body of still water to collect solar
energy and stores it as thermal energy, which is suitable for a variety of
applications, including electric power generation, industrial process
heating, and space conditioning.

As a practical matter, all ponds, lakes, oceans, and other expanses of
water in nature collect solar energy and convert it to thermal energy,
but their heat-retention efficiency is poor. This is because the water
near the surface is quickly cooled as the heat is rapidly dissipated to the
environment. The warmer and more buoyant water in the lower region
rises to the surface; this movement of water is called natural convection. Therefore, a water-pond system will be more effective in the
collection and storage of solar energy if convection is suppressed.
Several means of convection suppression have been suggested. By far
the most common approach is the establishment of a salt density gradient (increasing density with depth), so that the water in the lower
regions can be warmer than the water above it without simultaneously
acquiring lower density and rising to top by convection. Artificial salt-
gradient ponds were first investigated in Israel in the late 1950s and
have since been constructed as solar thermal energy sources in several
countries. They are usually referred to as "salt-gradient solar ponds"
or "solar ponds."

The mechanism of heat accumulation in a nonconvective solar pond
may be appreciated by comparing it with a normal convective pond.
Both the ponds absorb solar radiation in the water as well as in the
material at pond floor and convert it to heat. In the convective pond,

loss of heat is large because each small element of water, when heated, rises to the surface and exchanges heat with the atmosphere. In contrast, in the nonconvective pond, only the upper layer of water (at low temperature) can exchange heat with the atmosphere, with consequent low loss of heat. Heat loss from lower regions is by conduction only and is meager because nonconvective water is a poor conductor of heat. Thus, the regions of water near the bottom of the solar pond attain a high temperature, and if there is enough sunshine, this temperature can reach the boiling point while the pond surface remains close to that of ambient air.

II. HISTORICAL BACKGROUND

The physical phenomenon of the salt-gradient solar pond occurs in nature in some salt lakes which may have existed for a million years. The first documented description of these lakes was given by a Russian scientist (Von Kalecsinsky, 1902). He reported that during the summer the Madve Lagoon in Transylvania (Hungary) acquired temperatures in excess of 70°C at a depth of 1.32 m for a surface temperature close to that of ambient air. In more recent times several authors, including Anderson (1958), Wilson and Wellman (1962), Hoare (1966), Por (1970), Melack and Kilham (1972), Hudec and Sonnefeld (1974), and Cohen *et al.* (1977), have reported that salt lakes with elevated bottom-region temperatures occur in many parts of the world. The most remarkable seems to be Lake Vanda (Wilson and Wellman, 1962) in Antarctica. It exhibits a temperature of 25°C in the bottom region, while the annual mean atmospheric temperature at the lake site is -20°C and the lake is perennially covered with 3 to 4 m of ice. These lakes invariably have been observed to be characterized by salt leaching at the bottom of the lake and a supply of fresh water (low-salinity brine) at the surface provided by a river (or other means). The natural diffusion of salt gives rise to a downward increase in salt concentration which prevents convection and renders the upper region of lake a partially transparent insulator. Consequently, the lake acts as a solar heat trap and becomes heated in the lower region. Typical temperature and salt profiles for Lake Madve are illustrated in Fig. 1.

The idea of an artificial solar pond and its practical utilization as a solar thermal energy resource was proposed in Israel by Rudolph Bloch. The first laboratory-size solar pond was operated around 1959

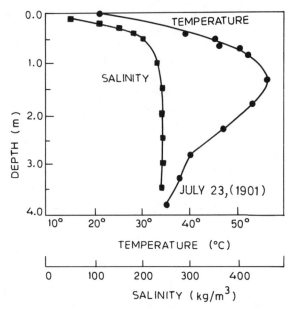

Fig. 1. Vertical profiles of temperature and salinity in Lake Madve. [After Kalec-sinsky (1902) and Hull (1979).]

(Tabor, 1959, 1961). Pioneering outdoor solar pond experiments were conducted by Tabor (1963) at Sdom on the shores of the Dead Sea; concentrated end brine primarily composed of magnesium chloride was used for generating the gradient. Subsequently, many investigations on the physics and engineering of the solar pond were reported (Weinberger, 1964; Elata and Levin, 1965; Tabor and Matz, 1965) and the technical feasibility of solar ponds was established. Temperatures up to 96°C were recorded, and the importance of careful site selection as well as the use of a liner to seal the pond's bottom and sides were stressed. Due to the massive thermal storage capacity of the system, the main emphasis was placed on the power production application. However, at that time, due to the availability of cheap fossil fuel-based electricity, such an application could not be demonstrated to be cost attractive. Solar pond research in Israel, therefore, suffered a setback after 1965, and for several years no significant work on solar ponds was reported in the literature.

The first solar pond outside Israel was built in Victoria (Australia) at Aspendale (Davey, 1968). Experimental solar ponds have since been constructed in several parts of the world. The countries making significant efforts on solar pond research are Australia (Davey, 1968; Sheridan, 1981; Collins, 1981, 1983; Golding *et al.*, 1983), Canada (Wilkins and Pinder, 1979), Chile (Hirschmann, 1965, 1970), India (Jain, 1973; Kaushika *et al.*, 1980; Bansal and Kaushika, 1981; Patel and Gupta, 1981; Kaushika and Bansal, 1982; Kaushika and Rao, 1983; Rao and Kaushika, 1983), Israel (Assaf, 1976; Tabor, 1975, 1980, 1981; Bronicki *et al.*, 1980, 1983; Tabor and Weinberger, 1980), Saudi Arabia (Nimmo *et al.*, 1981), the United States (Dake and Harleman, 1966; Huber and Harleman, 1968; Stolzenbach, 1968; Chepurniy and Savage, 1974; Rabl and Nielsen, 1975; Nielsen, 1975, 1976, 1980; Zangrando and Bryant, 1977, 1978; Badger *et al.*, 1977; Short *et al.*, 1978; Bryant *et al.*, 1979; Sargent, 1979; Multer, 1980; Wittenberg and Harris, 1981; Wittenberg and Etter, 1982), and the Soviet Union (Usmanov *et al.*, 1971, 1973a,b; Eliseev *et al.*, 1971, 1973). These investigations have explicity led to the identification of many solar pond applications, e.g., space heating, process heating, swimming pool heating, multistage desalination, and salt production. The world's record temperature for solar ponds has been registered as 109°C at about 2 m depth in the NaCl solar pond at the University of New Mexico on July 5, 1980 (Bryant, 1980). The boiling point of a saturated NaCl solution at sea level is 108.5°C, while that of a $MgCl_2$ solution is 125.5°C. Details about some experimental solar ponds are presented in Table I.

Besides the salt-gradient pond, several other solar pond concepts have also been tested or proposed (Sargent, 1979; Jayadev and Edesess, 1980). These are based on the premise that in a solar pond device the thermal losses to the environment are reduced either by suppressing the natural convection of water in the pond or by covering the pond surface with a partially transparent insulation. The concepts include the saturated (Roothmeyer, 1980) and unsaturated (Tabor, 1963) salt-gradient solar ponds, the gel or viscosity-stabilized pond (Shaffer, 1975, 1978; Kaushika *et al.*, 1982), the membrane pond (Rabl and Nielsen, 1975; Hull, 1980a; Kaushika *et al.*, 1980), the honeycomb stabilized solar pond (Ortabasi *et al.*, 1983), and the convective ponds (Dickinson *et al.*, 1976; Cassamajor and Parson, 1979).

The saturated salt-gradient pond is envisaged to be a stable and maintenance-free solar pond. It uses a salt whose solubility is a strongly increasing function of temperature; salts like sodium and potassium sulfate as well as potassium nitrate satisfy this condition. Salt is supplied at all depths in amounts large enough to maintain saturation

TABLE I

Salt-Gradient Solar Pond Experiments

Number	Location (reference)	Year	Salt	Area (m^2)	Depth (m)	Temperature reached (°C)
1	Sdom, Israel (Tabor, 1963)	1959	MgCl$_2$ (bittern)	625	1.0	96
2	Atlith, Israel (Tabor and Matz, 1965)	1964	MgCl$_2$ (bittern)	1375	1.5	74
3	Aspendale, Australia (Davey, 1968)	1964	NaCl	58	1.1	63
4	Bhavnagar, India (Jain, 1973)	1970	MgCl$_2$ (bittern)	1210	1.0	70
5	Columbus, Ohio (Nielsen, 1976)	1975	NaCl	200	2.5	62
6	Wooster, Ohio (Badger et al., 1977)	1975	NaCl	155	3.6	55
7	Albuquerque, New Mexico (Zangrando, 1979)	1975	NaCl	167	2.5	109 (in 1980)
8	Sdom, Israel (Sargent, 1979)	1975	MgCl$_2$	1100		103
9	Yavne, Israel (Tabor, 1981)	1977	NaCl	1500		90
10	Miamisburg, Ohio (Bryant et al., 1979; Wittenberg and Harris, 1981)	1978	NaCl	2000	3.0	50
11	Ein Bokek, Israel (Tabor, 1981)	1979	NaCl	7000	2.5	85
12	Alice Springs, Australia (Sheridan, 1981)	1981	NaCl	2000	2.2	70+
13	Montreal, Canada (Solar Pond, 1982)	1982	NaCl	700	2.0	70 (expected maximum)
14	Beth Ha'arva, Israel (Bronicki et al., 1983)	1983	—	4 ha	2.5	86

at each depth. The pond is hotter at the bottom than at top, so progressively increasing amounts of salt are dissolved toward the bottom. Due to saturation at all depths, the vertical diffusion of salt is impeded and the density gradient becomes stable.

The viscosity-stabilized pond employs thickening or gelling agents to render the pond water nonconvective. In its present state of technology, this concept is not economically competitive with the salt-gradient pond. The membrane-stratified solar pond uses closely spaced transparent membranes to suppress natural convection in the top region of the pond. The configuration of membranes may be horizontal sheets, vertical sheets, vertical tubes, or square honeycombs as shown in Fig. 2. The technical feasibility and economic viability of this concept has not yet been established. The convecting pond is often of shallow depth and is called the shallow solar pond. It consists of a water-filled plastic bag or pillow with glazing at the top and a blackened bottom which overlies the foam insulation. Energy from the system is extracted by the transport of the storage medium (water) itself, which eliminates the temperature difference between the storage and transport fluid. These ponds have been proposed to supply low-temperature hot water (40–60°C) for industrial processes as well as for electricity generation.

The best-documented type of solar pond is the salt-gradient solar

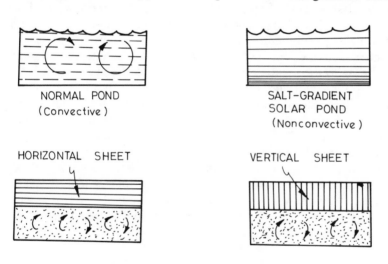

NORMAL POND
(Convective)

SALT–GRADIENT
SOLAR POND
(Nonconvective)

HORIZONTAL SHEET

VERTICAL SHEET

MEMBRANE–STRATIFIED PONDS

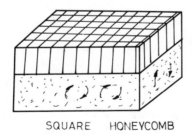

SQUARE HONEYCOMB

Fig. 2. Illustration of solar pond concepts.

pond. The next-best-understood is the shallow solar pond. The other ponds are still in the proof of concept stage. In this chapter solar ponds are, therefore, classified in three categories, which will be discussed in the following order: the salt-gradient solar pond (SGSP), the shallow solar pond (SSP), and alternative solar ponds (ASPs).

III. SALT-GRADIENT SOLAR PONDS: BASICS

A. Salt Selection

In the salt-gradient solar pond the density gradient is established by using dissolved salt. In principle, any inorganic salt can be used. However, some desirable characteristics for the usable salt are that it must be (1) cheap and readily available, (2) soluble in water without significant reduction of optical transmission, and (3) safe to handle and nonpolluting to the local air and hydrological environment. In most locations sodium chloride is the least expensive salt. But in some instances magnesium chloride, sodium sulfate, sodium carbonate, and other salts are also available at low cost as the waste products of mining or chemical processing. The cost and solubilities of some of these salts have been reported by Jayadev and Edesess (1980) and are shown in Fig. 3. Sodium chloride and magnesium chloride seem to satisfy all of the requirements for a usable salt and are thus the most widely used salts in the laboratory as well as in outdoor solar ponds. Most of the ponds in Israel have used bittern (magnesium chloride) for stabilization, while the ponds in the United States have used sodium chloride for stabilization.

B. Convective Stability

In the solar pond warmed by solar radiation, both the temperature T and the salt concentration S increase with depth. The density ρ_w of the pond fluid is, therefore, a function of salt concentration as well as of temperature. For the pond to be stable against natural convection, the magnitude of the density gradient on account of the salt concentration gradient must be greater than the negative density gradient produced by the temperature gradient; this is usually referred to as the static stability criterion. However, this criterion does not ensure stability against hydrodynamic instability processes in the solar pond, which

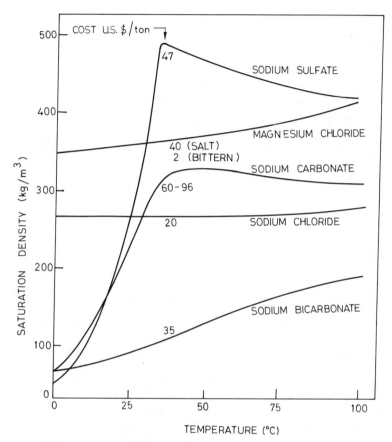

Fig. 3. Variation of saturation concentration with temperature for some salts. Çosts of salts in U.S. $ per ton are also indicated. [After Linke (1965) and Jaydev and Edesess (1980).]

have been mathematically investigated by Weinberger (1964), Veronis (1968), and Zangrando (1979). The resultant stability condition is referred to as the dynamic stability criterion, and for a solar pond with constant gradients of salt and temperature, it can be stated as

$$\frac{\partial S}{\partial x} \geq \left(\frac{P_r + 1}{P_r + \tau_S} \right) \frac{\alpha_c}{\beta_c} \frac{\partial T}{\partial x}, \tag{1}$$

where $P_r = \nu/\alpha_T$ is the Prandtl number and $\tau_S = \alpha_S/\alpha_T$ is the ratio of diffusivities; ν is kinematic viscosity; $\alpha_T = k_w/\rho_w c_w$ is the coefficient of temperature diffusivity; and α_S is the coefficient of salt diffusion. α_c and β_c are, respectively, the thermal and salt expansion coefficients.

Equation (3) may also be expressed as

$$\left(\frac{\partial S}{\partial x}\right)_{\min} = \frac{P_r + 1}{P_r + \tau_S} \frac{\alpha_c}{\beta_c} \frac{\partial T}{\partial x}. \tag{2}$$

Results of some sample calculations made from Eq. (2) for the minimum salt-gradient requirement corresponding to some typical surface and bottom characteristics of NaCl and MgCl$_2$ ponds are portrayed in Tables IIA and IIB, respectively.

C. Establishment and Maintenance of Gradient

Three methods have so far been proposed for the establishment of an initial salinity gradient. The first and the simplest method is to dump the total salt load into the pond half-filled with water; stirring is done to concentrate the solution. The upper half-portion is then filled with water. The upward diffusion of the salt establishes a density gradient which is stable in time if top and bottom concentrations are maintained constant by regularly washing the surface and adding salt to the bottom. This method has been referred to as "natural diffusion" (Zangrando, 1979). It is a very crude and slow method of establishing the salt gradient, and it should only be considered if the pond is very large or if the pond starting time could be unlimited.

The second method involves filling the pond with a number of layers of salt solutions of progressively decreasing concentration. The concentration of salt in successive layers is changed in steps from near saturation at the bottom to fresh water at the surface. This method is referred to as "stacking." The practical approach for stacking used in most solar ponds is to fill the bottom layer first and to float successively lighter layers on the lower and denser layers. However, some solar ponds in Australia have been built (Davey, 1968; Golding *et al.*, 1982) by successively feeding the solutions of increasing salinity into the bottom of the pond. For a typical pond of 1 m depth, one might use about 10 layers. The layering is accomplished with the help of a diffuser, illustrated in Fig. 4. It consists of two parallel plates separated by a small gap (10 mm). The brine is pumped through the center of the upper plate to exit horizontally at the perimeter.

TABLE IIA

Values of ($\partial S/\partial x$) Minimum for Different Surface and Bottom Characteristics for Sodium Chloride Solar Pond

Pond characteristics	Temperature (°C)	Salt concentration (kg/m³)	$\dfrac{\alpha_c}{\beta_c} = -\dfrac{\partial\rho/dT}{\partial\rho/dS}$ (kg/m³ °C)	$\dfrac{\nu + \alpha_T}{\nu + \alpha_S} \times \dfrac{\alpha_c}{\beta_c}$	($\partial S/\partial x$) Minimum		
					$\partial T/\partial x$ (100°C/m)	$\partial T/\partial x$ (250°C/m)	$\partial T/\partial x$ (500°C/m)
Pond surface	20	20	0.32	0.36	36	90	180
Cold bottom	20	260	0.65	0.71	71	177	354
Warm bottom	60	260	0.71	0.89	89	223	446
Hot bottom	100	260	0.84	1.18	118	295	590

TABLE IIB

Values of ($\partial S/\partial x$) Minimum for Different Surface and Bottom Characteristics for Magnesium Chloride Solar Pond

Pond characteristics	Temperature (°C)	Salt concentration (kg/m³)	$\dfrac{\alpha_c}{\beta_c} = -\dfrac{\partial\rho/dT}{\partial\rho/dS}$ (kg/m³ °C)	$\dfrac{\nu + \alpha_T}{\nu + \alpha_S} \times \dfrac{\alpha_c}{\beta_c}$	($\partial S/\partial x$) Minimum		
					$\partial T/\partial x$ (100°C/m)	$\partial T/\partial x$ (250°C/m)	$\partial T/\partial x$ (500°C/m)
Pond surface	20	20	0.27	0.30	30	75	150
Cold bottom	20	260	0.34	0.36	36	90	180
Warm bottom	60	260	0.48	0.54	54	135	270
Hot bottom	100	260	0.49	0.59	59	147.5	295

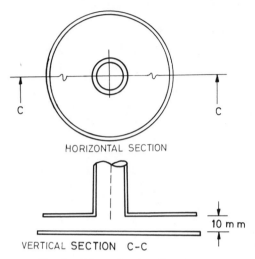

HORIZONTAL SECTION

VERTICAL SECTION   C-C

10 m m

Fig. 4.   Schematic of a diffuser.

The stepwise-type concentration profile is smoothed into a nearly linear concentration profile (Chepurniy and Savage, 1975) as a result of the diffusion process. The turbulent mixing generated during the filling process hastens the smoothing and tends to deplete the density gradient. This implies that there are limitations to the rate at which a solar pond can be filled. According to Tabor and Weinberger (1980), the brine exit velocity v at the diffuser perimeter can be expressed as follows.

(1)   When progressively lighter layers are stacked in a pond,

$$v_1 = \left[\alpha_t g x \frac{\rho_{n-1} - \rho_n}{\rho_n} \right]^{1/2}. \tag{3}$$

(2)   When successively denser layers are fed into the bottom,

$$v_2 = \left[\alpha_t g x \frac{\rho_{n-1} - \rho_n}{\rho_{n-1}} \right]^{1/2}, \tag{4}$$

and

$$v_2 = v_1 \rho_n / \rho_{n-1}, \tag{5}$$

where x, ρ_n, α_t are the thickness, density, and fraction of turbulent mixing in the nth layer. g is acceleration due to gravity. For typical pond layers with $x = 0.1$ m and $(\rho_{n-1} - \rho_n)/\rho_n = 0.03$, v_1 and v_2 are obtained as follows.

α	v_1 (m/sec)	v_2 (m/sec)
0.1	0.054	0.053
0.2	0.077	0.076
0.3	0.094	0.092
0.4	0.108	0.107
0.5	0.121	0.119

The third method is referred to as "redistribution." It is the most expedient method and may be adaptable to automatic controls. The method is based on the results of redistribution experiments (Nielsen and Rabl, 1976; Nielsen *et al.*, 1977; Zangrando, 1980) wherein it has been observed that when fresh water is injected at some level into homogeneous brine, it stirs and uniformly dilutes the brine from a few centimeters below the injection level to the surface. The procedure involves filling the pond with high-salinity brine to half of its total depth; fresh water is then fed into the pond through the diffuser. Initially the diffuser is placed at the bottom and the water is pumped through it to flow as an undercurrent in the pond. Owing to buoyancy the water flows to the top and stirs the intervening brine medium to uniform density. The brine surface level in the pond increases. The diffuser is then moved upward continuously or in steps; timings of the movement are so adjusted that the diffuser as well as the brine surface reach the final level at the same time. At the completion of the process, a uniform salt concentration gradient has been set up in the pond (Zangrando, 1980).

This vertical gradient of density causes an upward diffusion of salt in the pond. The rate of diffusion depends on salt diffusivity, density gradient, and wind-induced eddy diffusivity. For a typical sodium chloride solar pond, the amount of upward diffusion of salt is 60–80 gm/m^2 day (Savage, 1977; Tabor, 1980). A larger value of 200 gm/m^2 day has also been reported (Akbarzadah and Ahmadi, 1981). The tendency of salt diffusion is to destroy the gradient. One gradient control approach is to remove salt from the surface and inject it back into the bottom (Tabor, 1963). For this an evaporation pond would be required. Replenishment of the salt in the bottom can also be achieved by a passive technique (Akbarzadeh and Macdonald, 1982). However, a more prac-

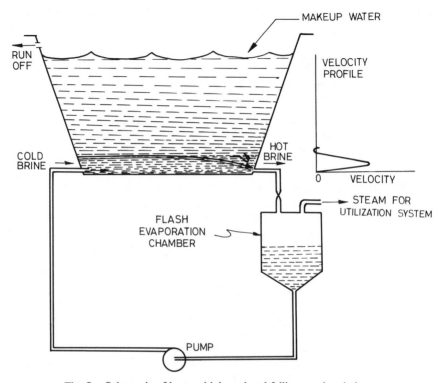

Fig. 5. Schematic of layer withdrawal and falling pond technique.

tical approach which does not require the continual addition of salt is the "falling pond" concept (Tabor, 1966, 1980; Shahar, 1968). A schematic of falling pond technique is shown in Fig. 5. Hot brine is withdrawn from the bottom layer without disturbing the layers above. This is possible since in a fluid system stratified with a density gradient, selective flow of the bottom layer can be accomplished without using a mechanical separation between the flowing and the stable regions of the system (Elata and Levin, 1962; Daniels and Merriom, 1975). The hot brine withdrawn from the solar pond is passed through the flash evaporator to remove some of its water. The solution then having a higher concentration and a smaller volume is reinjected into the pond bottom and the removed water is replaced into the surface layer. Consequently, the concentration gradient would be maintained. The fall in surface level due to evaporation is also restored by the addition of fresh water to keep both the pond depth and the surface concentration constant. In the preceding process the gradient tends to be displaced downward, hence the name "falling pond" technique. Following Tabor (1980), if v_0 is the vertical velocity (measured positively downward) of the mass of the brine in the pond; V_1 and V_2 are values of volume flow rates at input and output and A is the horizontal area of pond (assumed to be uniform vertically), then

$$v_0 A = V_2 - V_1.$$

The resultant downward flux of salt is given by

$$v_0 S = (V_2 - V_1)S/A.$$

The condition for the maintenance of the gradient may be written as

$$(V_2 - V_1)S/A = \alpha_S(\partial S/\partial x),$$

where α_S is coefficient of salt diffusion.

D. Absorption of Solar Radiation in the Pond

The pond surface is an air–water interface at which a small fraction of insolation is reflected back into space; the rest of the radiation enters the water and is attenuated (absorbed and scattered) throughout its depth and is almost completely absorbed at the black bottom. Some radiation may be reflected out of the water, but this amount is considered negligible in ponds.

The pond surface is always optically rough due to wind agitation. A satisfactory diffusive reflection theory of such a surface is not yet available, and pond surface characterization still relies on experimental results. Payne (1972) has observed that sea surface roughness caused by wind speeds up to 15 m/s had very little effect on surface reflectance. Therefore, the air–water interface at the pond surface may be assumed as planar and its optical reflection–transmission properties can be estimated from Fresnel's equations of classic electromagnetic theory, which are given next.

For direct (beam) radiation, the ray angle of incidence is equal to the solar zenith angle z (90 − altitude), which varies with time of the day, day of the year, and location; it is generally tabulated in meteorological tables as a function of time and latitude. If μ_w is the refractive index of water, then the ray angle of refraction θ_r is given by Snell's law,

$$\sin \theta_r = \sin z/\mu_w. \tag{6}$$

The transmission coefficient for an incident plane wave component polarized with its electric field vector parallel to the plane of incidence is given by

$$\tau_{\parallel} = \frac{4\mu_w \cos z \cos \theta_r}{(\mu_w \cos \theta_r + \cos z)^2},$$

and the transmission coefficient for the wave component polarized with its electric field vector normal to the plane of incidence is given by

$$\tau_{\perp} = \frac{4\mu_w \cos z \cos \theta_r}{(\mu_w \cos z + \cos \theta_r)^2}.$$

Direct insolation at the pond surface is unpolarized and hence the total transmission coefficient may be obtained by combining the components. Therefore we have

$$\tau = 2\mu_w(a^2 + b^2) \cos z \cos \theta_r, \tag{7}$$

where

$$a = 1/(\cos \theta_r + \mu_w \cos z) \tag{8a}$$

and

$$b = 1/(\cos z + \mu_w \cos \theta_r). \tag{8b}$$

Tabor and Weinberger (1980) have calculated the values of τ for direct radiation. In these calculations, an average angular position of the sun during the day is used and the refractive index of saline water in the pond is taken to be 1.33. Some results of their calculations are shown in Table III.

Diffuse sky radiation results from the scattering of both direct and ground reflected solar radiation by clouds and dust particles in the

TABLE III

Annual Mean Values of Transmission Coefficient τ at Air–Water Interface for Direct Radiation

		Latitude			
0	10	20	30	40	50
0.97	0.97	0.96	0.95	0.94	0.89

atmosphere. It accounts for approximately 15% of the total radiation when the sun is near the zenith and about 40% of the total radiation when the sun is near the horizon. Weinberger (1964) assumed the sky to be of uniform brightness and found the fraction of diffuse radiation penetrating the pond surface as

$$\tau_d = 2 \int_0^{\pi/2} \tau \sin z \cos z \, dz; \tag{9}$$

for low latitudes (<40°), $\tau_d \simeq 0.93$.

Experimentally observed attenuation of solar radiation in saline water is shown in Table IV. The short-wavelength portion of solar radiation penetrates several meters, whereas the near infrared is absorbed within a few centimeters. Several workers (Chepurniy and Savage, 1974; Bryant and Colbeck, 1977; Rabl and Nielsen, 1975; Kaushik *et al.*, 1980) have represented this attenuation by different empirical expressions. In Fig. 6, a comparative study of these expressions is made, and it is obvious that the superposition of five exponentials gives an excellent approximation of the intensity of radiation at different depths. Thus the intensity of solar radiation at any depth x and time t may be expressed as

$$S(x, t) = S(x = 0, t) \sum_{j=1}^{5} \mu j \exp(-n_j x),$$

TABLE IV

Energy Distribution (in Percent) in the Spectrum of Sunlight after Passing through Water Layers of Different Thicknesses[a]

Wavelength (μm)	Depth of water (m)				
	0	0.01	0.1	1	10
0.2–0.6	23.7	23.7	23.6	22.9	17.2
0.6–0.9	36.0	35.3	30.5	12.9	0.9
0.9–1.2	17.9	12.3	0.8	—	—
Over 1.2	22.4	1.7	—	—	—
Total	100.0	73.0	54.9	35.8	18.1

[a] After Rabl and Nielsen (1975) and Defant (1961).

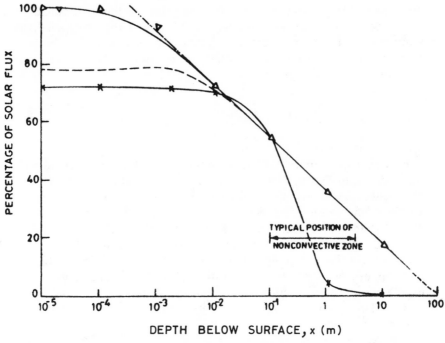

Fig. 6. Solar-radiant flux variation with depth below pond surface. ——, Observed (A. Defant, 1961); —·—, $0.36 - 0.08 \ln x$ (Bryant and Colbeck, 1971); —×—, $0.7239 \exp(-2.81x)$ (Chepurnly and Savage, 1974); ---, $\Sigma_{j=1}^{4} \mu j e^{-njx}$ (Rabl and Nielsen, 1975); —Δ—, $\Sigma_{j=1}^{5} \mu j e^{-njx}$ (Kaushika *et al.*, 1980). [After Kaushika *et al.* (1980). Copyright 1984, Applied Science Publishers Ltd.]

where $S(x = 0, t) = \tau S'(x = 0, t)$,

$$\mu_1 = 0.237, \quad n_1 = 0.032 \quad \text{m}^{-1},$$

$$\mu_2 = 0.193, \quad n_2 = 0.450 \quad \text{m}^{-1},$$

$$\mu_3 = 0.167, \quad n_3 = 3.000 \quad \text{m}^{-1},$$

$$\mu_4 = 0.179, \quad n_4 = 35.00 \quad \text{m}^{-1},$$

$$\mu_5 = 0.224, \quad n_5 = 255.0 \quad \text{m}^{-1}.$$

The expression due to Bryant and Colbeck (1977) is also a good approximation except in the region near $x = 0$. Therefore it should be used in the form

$$S(x, t) = S(x = 0, t)(0.36 - 0.08 \log x) \quad \text{(for } x > 0),$$

$$S(x, t) = S(x = 0, t) \quad \text{(for } x = 0).$$

(10)

E. Three Zones

The absorption of solar energy in the saline pond gives rise to a vertical temperature gradient which in combination with the existing salt gradient causes an upward diffusion of salt and heat. The pond is also in continuous interaction with all the atmospheric factors which act to induce a variety of physical and thermal processes in the pond-related environment. The major energy exchange mechanisms involved in these processes are (1) energy exchange in the saline solar pond: salt and heat diffusion in the gradient region; and (2) energy exchange between the pond and the environment: in-ground heat and mass transfer, wind effects and convection, convective and radiative surface cooling, and surface evaporation and rainfall.

Absorption of solar radiation in the pond water gives rise to a distributed heat source and the absorption of radiation by the blackened pond floor heats the bottom. Consequently, a nonuniform temperature gradient is set up in the vertical direction. Temperature gradients are relatively large in the surface region, where most of the infrared portion of the solar spectrum is absorbed, and in bottom region, where up to 30% of the radiation is absorbed by the pond floor. In solar ponds situated over low latitudes these gradients may reach a maximum of about 500°C/m in the surface region and about 400°C/m in the bottom region when no heat is extracted (Tabor and Weinberger, 1980).

Large temperature gradients in the bottom region induce thermohydrodynamic instabilities which make the bottom region convective and isothermal (Leshuk *et al.*, 1978; Nielsen, 1978a, 1980). The gradient in the bottom zone will tend to decrease with an increase in the thickness of the bottom convective zone. The equilibrium thickness of the bottom convective zone will obviously depend upon the amplitude of variation in solar radiation reaching the pond floor, the salinity gradient at the bottom of the pond, and the thermophysical properties of the supporting earthen regions. The bottom convective zone adds to the pond's stability by reducing the gradient of the bottom region temperature; it also provides thermal storage. A convective storage layer is, therefore, always placed under the gradient region in all modern solar ponds.

At the surface of the pond large temperature gradients result in the formation of a surface convective zone (Katti and Kaushika, 1981). Wind-generated surface currents and waves also contribute in the formation of this zone. Thus a salt-gradient solar pond normally consists of three zones: a convective zone at the surface, a second convective zone in the bottom, and a nonconvective zone between them.

Maintenance of appropriate thickness of the gradient zone is of basic importance for solar pond operation. The convective–nonconvective zone boundaries are the horizontal transition layers at which no fluid instabilities occur; thus, the average position of this boundary is expected to be stable against fluid-dynamic instabilities. However, in experimental ponds (Nielsen and Rabl, 1975; Nielsen, 1976), the boundary has been observed to move slowly, at typical speeds of the order of a few centimeters per month. This indicates that the steady-state boundary position is determined by processes much slower than the growth of thermohydrodynamic instabilities in the pond. The investigations of Elata and Levin (1965) and Nielsen (1978a,b, 1980) have enabled the identification of various physical processes involved in the determination of the zone boundary. Nielsen (1978a), Gupta and Patel (1979), and Kamal and Nielsen (1983) have suggested that this boundary is determined by a balance between convective erosion of the gradient zone and salt diffusion enlargement of the same gradient zone and that the required boundary concentration gradient G_c may be characterized in terms of the boundary temperature gradient as

$$\log G_c = 1.36 + 0.66 \log G_T$$

F. Thermal Collection Efficiency

The three-zone solar pond has been analyzed as a steady-state flat-plate solar-energy collector by Kooi (1979, 1981), Hawlader (1980), Nielsen (1980), and Bansal and Kaushika (1981). Thermal efficiency of about 25% has been predicted at a collection temperature of 95°C. Zangrando and Bryant (1977) have indicated that the total heat accumulated in the bottom convective zone is 25% of insolation; of this about 5% is stored in the ground and 5% is lost through the sidewalls. Thus the collection efficiency of the solar pond is between 15 and 25%. This is rather low, but it is more than compensated by the fact that the solar pond combines in itself both collection and storage and it continuously collects the same fraction of incident energy irrespective of insolation and atmospheric temperature.

IV. THERMAL MODELS OF THE SALT-GRADIENT POND

A. General Description

The purpose of a thermal model is engineering analysis or design. The nature of the problem involved is that of calculating temperatures in a pond environment wherein heat transfer is taking place. The temperature distribution may itself be of primary interest, or it may be used to calculate the heat fluxes and the efficiency of heat collection in the system.

In practical solar ponds the lateral dimensions are large compared to their depths, and hence the heat flow is essentially vertical. The mathematical model of such a configuration was first given by Weinberger (1964). The approach consisted of solving the one-dimensional Fourier heat conduction equation with a heat source term (due to volume absorption of solar radiation) by using appropriate boundary conditions at the surface and bottom of the pond. Subsequently, Stolzenbach *et al.* (1968), Eliseev *et al.* (1971), Dake (1973), Chepurniy and Savage (1974), and Akbarzadeh and Ahmadi (1980) have followed the same approach to investigate more realistic physical situations. Both analytical and numerical solutions corresponding to periodic and arbitrary transient boundary conditions have been obtained. All these analyses are based on a solar pond model which assumes that the pond water behaves as a single nonconvective zone. As a practical matter, the solar pond is a three-zone configuration. Rabl and Nielsen (1975) suggested the installation of a transparent cover at the surface to suppress the convective zone; these authors also gave an analytical treatment of the resultant two-zone solar pond, which was subsequently modified for more realistic physical situations by Kaushika *et al.* (1980). The three-zone configuration of the solar pond has also been analyzed; the time-dependent closed form analytical solution (Sodha *et al.* 1981; Kaushika and Rao, 1983; Rao, 1983), as well as the numerical finite difference solution (Hawlader and Brinkworth, 1981; Sheridan, 1982) corresponding to periodic and arbitrary transient boundary conditions, have been obtained. In what follows a discussion of thermal models of three-zone solar pond is presented; the density ρ_w, the specific heat C_w, and the thermal conductivity K_w of brine are assumed to be independent of temperature and salinity at all depths. It is also assumed that the convective–nonconvective zone boundaries are stationary in time.

B. Formulation of Thermal Transfers

A schematic diagram of a three-zone solar pond with stationary boundaries is shown in Fig. 7. Following Sodha et al. (1981), Kaushika and Rao (1983), and Rao (1983), the equations governing the temperature and heat flux in various zones and their interfaces are given as the following.

1. Nonconvective Zone

The Fourier heat conduction equation for this zone contains a source term arising due to absorption of solar radiation throughout the pond. It is given by

$$K_w \frac{\partial^2 T(x, t)}{\partial x^2} = \rho_w C_w \frac{\partial T(x, t)}{\partial t} + \frac{\partial S(x, t)}{\partial x}. \tag{11}$$

The boundary conditions are at $x = l_1$

$$-K_w \frac{\partial T(x, t)}{\partial x}\bigg|_{x=l_1} = h_1[T_1(t) - T(x = l_1, t)] \tag{12}$$

and at $x = l_1 + l_2$

$$-K_w \frac{\partial T(x, t)}{\partial x}\bigg|_{x=l_1+l_2} = h_2[T(x = l_1 + l_2, t) - T_w(t)]. \tag{13}$$

2. Ground Zone

The heat conduction equation for the ground zone is

$$K_g \frac{\partial^2 T(x, t)}{\partial x^2} = \rho_g C_g \frac{\partial T(x, t)}{\partial t}. \tag{14}$$

The boundary conditions are

$$-K_g \frac{\partial T(x, t)}{\partial x}\bigg|_{x=l_1+l_2+l_3}$$

$$= h_3[T_w(t) - T(x = l_1 + l_2 + l_3, t)] + \alpha S(x = l_1 + l_2 + l_3, t)$$

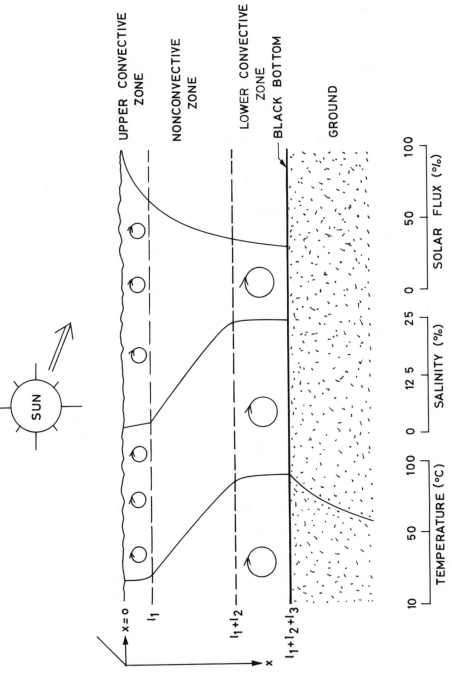

Fig. 7. Three-zone solar pond.

or

$$-K_g \left.\frac{\partial T(x, t)}{\partial x}\right|_{x=l_1+l_2+l_3} = h_3[T'_w(t) - T(x = l_1 + l_2 + l_3, t)], \quad (15)$$

where

$$T'_w(t) = T_w + (\alpha/h_3)S(x = l_1 + l_2 + l_3, t)$$

and

$$T(x, t) \text{ is finite as } x \to \infty. \quad (16)$$

3. Upper Convective Zone: Energy Balance

It is assumed that this zone has uniform temperature $T_1(t)$, which may be expressed in terms of atmospheric air temperature $T_a(t)$ by the consideration of energy balance of this zone, i.e.,

$$\begin{bmatrix} \text{Solar heat flux} \\ \text{absorbed between} \\ x = 0 \text{ and } x = l_1 \end{bmatrix} = \begin{bmatrix} \text{Heat loss to} \\ \text{the atmosphere} \\ \text{due to radiation} \\ \text{and convection} \end{bmatrix} + \begin{bmatrix} \text{Heat loss to} \\ \text{the atmosphere} \\ \text{by vaporization} \end{bmatrix}$$

$$+ \begin{bmatrix} \text{Heat conducted to} \\ \text{the nonconvective} \\ \text{zone through the} \\ \text{interface at } x = l_1 \end{bmatrix}$$

$$+ \begin{bmatrix} \text{The amount of} \\ \text{heat stored in} \\ \text{the convective} \\ \text{zone} \end{bmatrix}, \quad (17)$$

$$\Delta S(x, t) = h_0[T_1(t) - T_a(t)] + \dot{Q}_e + h_1[T_1(t) - T(x = l_1, t)]$$
$$+ z_w l_1[dT_1(t)/dt], \quad (18)$$

where

$$\Delta S(x, t) = S(x = 0, t) - S(x = l_1, t).$$

The heat loss due to evaporation from the surface $\dot{Q}_e$ may be given by the expression

$$\dot{Q}_e = h_e[p_s(T_1(t)) - \gamma p_s[T_a(t)], \quad (19)$$

where $p_s[T_1(t)]$ and $p_s[T_a(t)]$ are saturated vapour pressures at temperatures $T_1(t)$ and $T_a(t)$, respectively.

In general the saturated vapor pressure is a nonlinear function of temperature; however, since the range of the annual cycle of variation in $T_a(t)$ as well as in $T_1(t)$ is rather small, $p_s(T)$ may be expressed as a linear function

$$p_s(T) = C_1 T - C_2'. \tag{20}$$

Combining Eqs. (19) and (20) we have

$$\dot{Q}_e = h_e\{[T_1(t) - \gamma T_a(t)]C_1 - C_2\}, \tag{21}$$

where $C_2 = C_2'(1 - \gamma)$.

4. Lower Convective Zone

The lower zone is also a convective zone and has uniform temperature $T_w(t)$ throughout its depth. The energy balance in this zone is given by

$$z_w l_3 \frac{dT_w(t)}{dt} = -K_w \frac{\partial T(x, t)}{\partial x}\bigg|_{x=l_1+l_2}$$

$$+ (1 - \alpha)S(x = l_1 + l_2 + l_3, t)$$

$$+ S(x = l_1 + l_2, t) - S(x = l_1 + l_2 + l_3, t)$$

$$- h_3[T_w(t) - T(x = l_1 + l_2 + l_3, t)] - \dot{Q}(t). \tag{22}$$

The boundary condition at $x = l_1 + l_2 + l_3$ is

$$-K_g \frac{\partial T(x, t)}{\partial x}\bigg|_{x=l_1+l_2+l_3} = h_3[T_w(t) - T(x = l_1 + l_2 + l_3, t)]$$

$$+ \alpha S(x = l_1 + l_2 + l_3, t). \tag{23}$$

Combining Eqs. (22) and (23) we have

$$\dot{Q}(t) = -K_w \frac{\partial T(x, t)}{\partial x}\bigg|_{x=l_1+l_2} + K_g \frac{\partial T(x, t)}{\partial x}\bigg|_{x=l_1+l_2+l_3}$$

$$+ S(x = l_1 + l_2, t) - z_w l_3 \frac{dT_w(t)}{dt}. \tag{24}$$

In the preceding formulation it is assumed that the radiations reflected (solar spectrum) and emitted (infrared) by the pond bottom are completely absorbed within the bottom convective zone.

The first and second terms on the right-hand side of Eq. (24) may be evaluated by solving the appropriate Fourier heat conduction equations with boundary conditions as follows.

C. Analytical Model*

An analytical model of a solar pond due to Sodha *et al.* (1981), Kaushika and Rao (1983), and Rao (1983) involves the periodic solution of heat conduction equations. The model has special merit on account of its mathematical elegance, its closed form solution, and its capacity to enable system optimization.

1. Periodic Solution of Heat Conduction Equations

The solar pond provides long-term storage spanning seasons; so in an evaluation of its yearround thermal performance one need not consider the hourly and daily fluctuations of insolation $S'(x = 0, t)$ and the ambient air temperature $T_a(t)$. The observed annual variations in $S'(x = 0, t)$ and $T_a(t)$ may be regarded as periodic in nature and represented by a Fourier series; six harmonics are found to provide a sufficiently good representation. Therefore we have

$$S'(x = 0, t) = S'_{00} + \sum_{m=1}^{6} S'_{0m} \exp(im\omega t). \tag{25}$$

Therefore

$$S(x, t) = \left[S_{00} + \sum_{m=1}^{6} S_{0m} \exp(im\omega t) \right] \left[\sum_{j=1}^{5} \mu_j \exp(-n_j x) \right], \tag{26}$$

$$T_a(t) = T_{a0} + \sum_{m=1}^{6} T_{am} \exp(im\omega t),$$

where

$$\omega = 2\pi/(1 \times 365.25 \times 24 \times 60 \times 60) \quad \text{s}^{-1}.$$

Due to the periodic nature of solar intensity and ambient air temperature as expressed in Eqs. (25) and (26), $T_1(t)$, $T_w(t)$, $T'_w(t)$, and $T(x, t)$ will also be periodic in nature and hence may be expressed as

$$T_1(t) = T_{10} + \sum_{m=1}^{6} T_{1m} \exp(im\omega t), \tag{27a}$$

$$T_w(t) = T_{w0} + \sum_{m=1}^{6} T_{wm} \exp(im\omega t), \tag{27b}$$

* Discussion closely follows Rao (1983).

$$T'_w(t) = T'_{w0} + \sum_{m=1}^{6} T'_{wm} \exp(im\omega t), \qquad (27c)$$

$$T(x, t) = T_0(x) + \sum_{m=1}^{6} T_m(x) \exp(im\omega t). \qquad (27d)$$

Substituting the value of $\dot{Q}_e$ from Eq. (21) in Eq. (18) and equating time-independent and time-dependent parts, expressions for T_{10} and T_{1m} are obtained as

$$T_{10} = \frac{\Delta S_0 + T_{a0}(h_0 + h_e \gamma c_1) + h_1 T_0(x = l_1) + h_e c_2}{h_0 + h_e c_1 + h_1}, \qquad (28)$$

$$T_{1m} = \frac{\Delta S_m + T_{am}(h_0 + h_e \gamma c_1) + h_1 T_m(x = l_1)}{h_0 + h_e c_1 + h_1 + im\omega Z_w l_1}, \qquad (29)$$

where

$$\Delta S_0 = S_{00} \sum_{j=1}^{5} \mu_j [1 - \exp(-n_j l_1)]$$

and

$$\Delta S_m = S_{0m} \sum_{j=1}^{5} \mu_j [1 - \exp(-n_j l_1)].$$

Now, substituting the value of $T_1(t)$ in Eq. (12) and characterizing the boundary condition at $x = l_1$ and separating the time-independent and time-dependent parts, we have

$$-K_w \left. \frac{\partial T_0(x)}{\partial x} \right|_{x=l_1} = h'_{10}[T'_{a0} - T_0(x = l_1)], \qquad (30a)$$

$$-K_w \left. \frac{\partial T_m(x)}{\partial x} \right|_{x=l_1} = h'_{1m}[T'_{am} - T_m(x = l_1)], \qquad (30b)$$

where

$$h'_{10} = \frac{h_1(h_0 + h_e c_1)}{h_0 + h_e c_1 + h_1},$$

$$h'_{1m} = \frac{h_1(h_0 + h_e c_1 + im\omega Z_w l_1)}{h_0 + h_e c_1 + h_1 + im\omega l_1},$$

$$T'_{a0} = \frac{\Delta S_0 + h_e c_2 + T_{a0}(h_0 + h_e \gamma c_1)}{h_0 + h_e c_1 + h_1},$$

and

$$T'_{am} = \frac{\Delta S_m + (h_0 + h_e\gamma c_1)T_{am}}{h_0 + h_e c_1 + h_1}.$$

Equation (11) is a nonhomogeneous differential equation, and its periodic solution with boundary conditions expressed by Eqs. (30) and (13) is

$$T(x, t) = -\sum_{j=1}^{5} \frac{\mu_j S_{00} \exp(-n_j x)}{n_j K_w} + d_1 x + d_2$$

$$+ \sum_{m=1}^{6} \left\{ [A_m \exp(\beta_m x) + B_m \exp(-\beta_m x)] \right.$$

$$\left. + \sum_{j=1}^{5} \frac{\mu_j n_j \exp(-n_j x)}{K_w(\beta_m^2 - n_j^2)} \right\} S_{0m} \exp(im\omega t), \qquad (31)$$

where

$$\beta_m = (im\omega\rho_w C_w/K_w)^{1/2},$$

$$d_1 = \left\{ (T_{w0} - T'_{a0}) - \sum_{j=1}^{5} S_{00}\mu_j \exp(-n_j l_1) \left[\left(\frac{1}{h'_{10}} + \frac{1}{n_j K_w} \right) \right. \right.$$

$$\left. + \left(\frac{1}{h_2} - \frac{1}{n_j K_w} \right) \exp(-n_j l_2) \right] \right\}$$

$$\div \left[K_w \left(\frac{1}{h'_{10}} + \frac{1}{h_2} + \frac{l_2}{K_w} \right) \right], \qquad (32a)$$

$$d_2 = \left\{ T'_{a0} \left(\frac{1}{h_2} + \frac{l_1 + l_2}{K_w} \right) - T_{w0} \left(\frac{l_1}{K_w} - \frac{1}{h'_{10}} \right) \right.$$

$$+ S_{00} \sum_{j=1}^{5} \mu_j \exp(-n_j l_1) \left[\left(\frac{1}{h_2} + \frac{l_1 + l_2}{K_w} \right) \right.$$

$$\left(\frac{1}{h'_{10}} + \frac{1}{n_j K_w} \right)$$

$$\left. \left. + \left(\frac{l_1}{K_w} - \frac{1}{h'_{10}} \right) \left(\frac{1}{h_2} - \frac{1}{n_j K_w} \right) \exp(-n_j l_2) \right] \right\}$$

$$\div \left[\frac{1}{h'_{10}} + \frac{1}{h_2} + \frac{l_2}{K_w} \right], \qquad (32b)$$

$$A_m = \frac{1}{D_{1m}} \left(\sum_{j=1}^{5} \frac{\mu_j n_j S_{0m}}{K_w \beta_m (\beta_m^2 - n_j^2)} \left\{ \left(\frac{1}{h'_{1m}} + \frac{1}{K_w \beta_m} \right) \right. \right.$$

$$\times \left(\frac{n_j}{h_2} - \frac{1}{K_w} \right) \exp[-n_j(l_1 + l_2)] - \left(\frac{1}{h_2} - \frac{1}{K_w \beta_m} \right)$$

$$\times \left(\frac{n_j}{h'_{1m}} + \frac{1}{K_w} \right) \exp(-n_j l_1) \exp(-\beta_m l_2) \Big\}$$

$$+ \frac{1}{K_w \beta_m} \left[\left(\frac{1}{h_2} - \frac{1}{K_w \beta_m} \right) \right.$$

$$\left. \left. \times \exp(-\beta_m l_2) T'_{am} + \left(\frac{1}{h'_{1m}} + \frac{1}{K_w \beta_m} \right) T_{wm} \right] \right), \qquad (32c)$$

$$B_m = \frac{1}{D_{2m}} \left(\sum_{j=1}^{5} \frac{\mu_j n_j S_{0m}}{K_w \beta_m (\beta_m^2 - n_j^2)} \left\{ \left(\frac{1}{h'_{1m}} - \frac{1}{K_w \beta_m} \right) \right. \right.$$

$$\times \left(\frac{n_j}{h_2} - \frac{1}{K_w} \right) \exp[-n_j(l_1 + l_2)] - \left(\frac{1}{h_2} + \frac{1}{K_w \beta_m} \right)$$

$$\times \left(\frac{n_j}{h'_{1m}} + \frac{1}{K_w} \right) \exp(-n_j l_1) \exp(\beta_m l_2) \Big\}$$

$$+ \frac{1}{K_w \beta_m} \left[\left(\frac{1}{h_2} + \frac{1}{K_w \beta_m} \right) \exp(\beta_m l_2) T'_{am} \right.$$

$$\left. \left. + \left(\frac{1}{h'_{1m}} - \frac{1}{K_w \beta_m} \right) T_{wm} \right] \right). \qquad (32d)$$

From Eq. (31) we have

$$-K_w \frac{\partial T}{\partial x} (x, t) \Big|_{x=l_1+l_2}$$

$$= -S_{00} \sum_{j=1}^{5} \mu_j \exp[-n_j(l_1 + l_2)]$$

$$-K_w d_1 - K_w \left[\sum_{m=1}^{6} \left(\beta_m \{ A_m \exp[\beta_m(l_1 + l_2)] \right. \right.$$

$$- B_m \exp[-\beta_m(l_1 + l_2)] \}$$

$$\left. \left. -S_{0m} \sum_{j=1}^{5} \frac{\mu_j n_j^2 \exp[-n_j(l_1 + l_2)]}{K_w \beta_m (\beta_m^2 - n_j^2)} \right) \exp(im\omega t) \right]. \qquad (33)$$

Substituting the value of d_1 from Eq. (32a) in Eq. (33) we have

$$-K_w \frac{\partial T(x, t)}{\partial x}\bigg|_{x=l_1+l_2} = -S_{00} \sum_{j=1}^{5} \mu_j \exp[-n_j(l_1 + l_2)]$$

$$+ [(\alpha\tau)_{\text{eff}} S_{00} - U_L(T_{w0} - T'_{a0})]$$

$$- K_w \bigg\{ \sum_{m=1}^{6} \bigg[\beta_m \Big(\{A_m \exp[\beta_m(l_1 + l_2)]$$

$$- B_m \exp[-\beta_m(l_1 + l_2)]\}$$

$$- \sum_{j=1}^{5} \frac{S_{0m}\mu_j n_j^2 \exp[-n_j(l_1 + l_2)]}{K_w \beta_m(\beta_m^2 - n_j^2)} \Big) \bigg] \exp(im\omega t) \bigg\},$$

$$(34)$$

where U_L and $(\alpha\tau)_{\text{eff}}$ are expressed as

$$\frac{1}{U_L} = \frac{1}{h'_{10}} + \frac{1}{h_2} + \frac{l_2}{K_w},$$

$$(\alpha\tau)_{\text{eff}} = U_L \sum_{j=1}^{5} \bigg[\Big(\frac{1}{h'_{10}} + \frac{1}{n_j K_w} \Big)$$

$$+ \Big(\frac{1}{h_2} - \frac{1}{n_j K_w} \Big) \exp(-n_j l_2) \bigg] \mu_j \exp(-n_j l_1).$$

Similarly, the solution of heat conduction equation (14) is given by

$$T(x, t) = A_0 x + B_0 + \sum_{m=1}^{6} [A'_m \exp(\beta_{mg}x)$$

$$+ B'_m \exp(-\beta_{mg}x)] \exp(im\omega t). \quad (35)$$

The preceding equation with boundary conditions given by Eqs. (15) and (16) leads to

$$-K_g \frac{\partial T(x, t)}{\partial x}\bigg|_{x=l_1+l_2+l_3} = \sum_{m=1}^{6} \frac{T'_{wm} \exp(im\omega t)}{1/h_3 + 1/K_g\beta_{mg}}, \quad (36)$$

where

$$\beta_{mg} = [(im\omega\rho_g C_g)/K_g]^{1/2}.$$

Furthermore, we have

$$S(x = l_1 + l_2, t) = \left[S_{00} + \sum_{m=1}^{6} S_{0m} \exp(im\omega t) \right]$$

$$\times \left\{ \sum_{j=1}^{5} \mu_j \exp[-n_j(l_1 + l_2)] \right\}, \qquad (37)$$

$$Z_w l_3 \frac{dT_w(t)}{dt} = im\omega Z_w l_3 T_{wm} \exp(im\omega t). \qquad (38)$$

Thus the rate of heat retrieval from the solar pond may be expressed by using Eqs. (24) and (34)–(38) as

$$\dot{Q}(t) = [S_{00}(\alpha\tau)_{\text{eff}} - U_L(T_{w0} - T'_{a0})]$$

$$+ \sum_{m=1}^{6} \left\{ S_{0m} \sum_{j=1}^{5} \left\{ \mu_j \exp[-n_j(l_1 + l_2)] \left(1 + \frac{n_j^2}{\beta_m^2 - n_j^2} \right) \right. \right.$$

$$- N_j \exp[\beta_m(l_1 + l_2)] + M_j \exp[-\beta_m(l_1 + l_2)]$$

$$\left. - \frac{\mu_j \exp[-n_j(l_1 + l_2 + l_3)]}{1/h_3 + 1/K_g\beta_{mg}} \right\} - T'_{am} \left\{ \left(\frac{1}{h_2} - \frac{1}{K_w\beta_m} \right) \right.$$

$$\times \frac{\exp(\beta_m l_1)}{D_{1m}} - \left(\frac{1}{h_2} + \frac{1}{K_w\beta_m} \right) \frac{\exp(-\beta_m l_1)}{D_{2m}} \right\}$$

$$- T_{wm} \left\{ \frac{1}{1/h_3 + 1/K_g\beta_{mg}} + \left(\frac{1}{h'_{1m}} + \frac{1}{K_w\beta_m} \right) \right.$$

$$\exp[\beta_m(l_1 + l_2)] \frac{1}{D_{1m}}$$

$$\left. - \left(\frac{1}{h'_m} - \frac{1}{K_w\beta_m} \right) \frac{\exp[-\beta_m(l_1 + l_2)]}{D_{2m}} \right\}$$

$$+ Z_w l_3 im\omega \right\} \exp(im\omega t), \qquad (39)$$

where

$$D_{1m} = \left(\frac{1}{h'_{1m}} + \frac{1}{K_w\beta_m} \right) \left(\frac{1}{h_2} + \frac{1}{K_w\beta_m} \right) \exp[\beta_m(l_1 + l_2)]$$

$$- \left(\frac{1}{h'_{1m}} - \frac{1}{K_w\beta_m} \right) \left(\frac{1}{h_2} - \frac{1}{K_w\beta_m} \right) \exp[\beta_m(l_1 - l_2)],$$

$$D_{2m} = \left(\frac{1}{h'_{1m}} + \frac{1}{K_w\beta_m}\right)\left(\frac{1}{h_2} + \frac{1}{K_w\beta_m}\right)\exp[-\beta_m(l_1 - l_2)]$$

$$- \left(\frac{1}{h'_{1m}} - \frac{1}{K_w\beta_m}\right)\left(\frac{1}{h_2} - \frac{1}{K_w\beta_m}\right)\exp[-\beta_m(l_1 + l_2)],$$

$$N_j = \frac{1}{D_{1m}}\frac{\mu_j n_j}{\beta_m^2 - n_j^2}\left\{\left(\frac{1}{h'_{1m}} + \frac{1}{K_w\beta_m}\right)\left(\frac{n_j}{h_2} - \frac{1}{K_w}\right)\right.$$

$$\times \exp[-n_j(l_1 + l_2)] - \left(\frac{1}{h_2} - \frac{1}{K_w\beta_m}\right)\left(\frac{n_j}{h'_{1m}} + \frac{1}{K_w}\right)$$

$$\left. \exp(-n_j l_1)\exp(-\beta_m l_2)\right\},$$

$$M_j = \frac{1}{D_{2m}}\frac{\mu_j n_j}{\beta_m^2 - n_j^2}\left[\left(\frac{1}{h'_{1m}} - \frac{1}{K_w\beta_m}\right)\left(\frac{n_j}{h_2} - \frac{1}{K_w}\right)\right.$$

$$\left. - \left(\frac{1}{h_2} + \frac{1}{K_w\beta_m}\right)\left(\frac{n_j}{h'_{1m}} + \frac{1}{K_w}\right)\exp(-n_j l_1)\exp(\beta_m l_2)\right].$$

2. Extraction of Heat

Solar energy is collected, concentrated, and stored in the lower convecting or storage zone and could be withdrawn as heat energy from this zone. There are two methods which have so far been employed for the withdrawal. In the first method heat exchange pipes are placed in the bottom region of the pond. The heat exchange fluid while flowing through these pipes gets heated and carries the heat to the system in which it is used. Hipsher and Boehm (1976), Jain (1973), Nielsen (1976, 1978, 1980), and Wittenberg and Harris (1981) have experimentally verified this method of heat extraction and found that the convective heat transfer around the heat exchanger is quite good. However, to obtain a small temperature difference between the pond and heat exchange fluid, a large surface area of pipe would be required and may prove to be expensive. In the second method high-density brine from the storage zone is extracted and sent to external heat exchangers, at which heat is removed and the cold brine is sent back to the pond bottom at an appropriate flow rate. Several workers (Tabor and Matz, 1965; Elata and Levin, 1965; Zangrando and Bryant, 1977; and Wittenberg and Etter, 1982) have experimentally tested this process of heat extraction and have found it to be the most appropriate technique for large-area solar ponds.

In both the processes heat can be extracted in two modes: (1) Heat extraction by circulating the heat exchange fluid at constant mass flow rate: This will cause the temperature of heat extraction zone as well as of the heat exchange fluid to vary with time. (2) Heat extraction by keeping the temperature of heat extraction zone as constant: This will obviously require the variation in flow rate of the heat exchange fluid. In the following we consider these two modes of heat extraction.

a. Constant Flow Heat Extraction. If the heat is extracted by flowing the heat exchange fluid, water, at constant flow rate, then

$$\dot{Q}(t) = (m_w C_w / A)[T_{out}(t) - T_i(t)].$$

If outlet temperature is taken as $T_w(t)$, we have

$$\dot{Q}(t) = GC_w[T_w(t) - T_i(t)].$$

Similarly, in case of heat removal using heat exchangers in the pond, following Rao and Kaushika (1983) and Rao (1983), we have

$$\dot{Q}(t) = GC_w F_R[T_w(t) - T_i(t)], \tag{40}$$

where

$$F_R = 1 - \exp[(-h_t \rho L / A)/GC_w].$$

The variation of the heat removal Factor F_R has been studied in detail by Rao and Kaushika (1983). It can be seen From Eq. (40) that heat extraction by the circulating brine method can be considered as a particular case of heat removal by a heat exchange pipe and it corresponds to $F_R = 1$.

Further, when the flow rate of the heat exchange fluid is maintained as constant, the temperature of the heat extraction zone $T_w(t)$ will vary. It can be evaluated by substituting the value of $\dot{Q}(t)$ from Eq. (39) in Eq. (40). Thus, we have

$$T_w(t) = \frac{S_{00}(\alpha\tau)_{eff} - U_L T'_{a0} + GC_w F_R T_{i0}}{U_L + GC_w F_R}$$

$$+ \sum_{m=1}^{6} \left\{ S_{0m} \sum_{j=1}^{5} \left[\left(1 + \frac{n_j^2}{\beta_m^2 - n_j^2} \right) \mu_j \exp[-n_j(l_1 + l_2)] \right. \right.$$

$$- \frac{\mu_j \exp[-n_j(l_1 + l_2 + l_3)]}{1/h_3 - 1/K_g \beta_{mg}} - N_j \exp[\beta_m(l_1 + l_2)]$$

$$\left. + M_j \exp[-\beta_m(l_1 + l_2)] \right\} - T'_{am} \left\{ \left(\frac{1}{h_2} - \frac{1}{K_w \beta_m} \right) \frac{\exp(\beta_m l_1)}{D_{1m}} \right.$$

$$- \left(\frac{1}{h_2} + \frac{1}{K_w \beta_m} \right) \frac{\exp(-\beta_m l_1)}{D_{2m}} \Big\}$$

$$+ GC_w F_R T_{im} \div \left[\left(\frac{1}{h'_{1m}} + \frac{1}{K_w \beta_m} \right) \frac{\exp[\beta_m(l_1 + l_2)]}{D_{1m}} \right.$$

$$- \left(\frac{1}{h_2} - \frac{1}{K_w \beta_m} \right) \frac{\exp[-\beta_m(l_1 + l_2)]}{D_{2m}} + Z_w im\omega l_3$$

$$+ \left. \frac{1}{1/h_3 + 1/K_g \beta_{mg}} + GC_w F_R \right] \Big\} \exp(im\omega t). \tag{41}$$

In the preceding formulation the inlet water temperature variation is also assumed to be represented by a Fourier series of six harmonics:

$$T_i(t) = T_{i0} + \sum_{m=1}^{6} T_{im} \exp(im\omega t). \tag{42}$$

b. Heat Extraction at Constant Temperature. The expression for a transient rate at which heat can be extracted from a three-zone solar pond at constant temperature of the heat extraction zone (bottom convective zone) can be obtained by substituting $T_{wm} = 0$ and $T_{w0} = T_{wc}$ in Eq. (39). Therefore we have

$$Q_c(t) = [S_{00}(\alpha\tau)_{\text{eff}} - U_L(T_{wc} - T'_{a0})$$

$$+ \sum_{m=1}^{6} \Big\{ S_{0m} \left[\sum_{j=1}^{5} \left(\frac{\mu_j \exp[-n_j(l_1 + l_2 + l_3)]}{1/h_3 + 1/K_g \beta_{mg}} \right. \right.$$

$$+ \left(1 + \frac{n_j^2}{\beta_m^2 - n_j^2} \right) \mu_j \exp[-n_j(l_1 + l_2)] - N_j \exp[\beta_m(l_1 + l_2)]$$

$$+ M_j \exp[-\beta_m(l_1 + l_2)] \Big) \Big] - T_{am} \left[\left(\frac{1}{h_2} - \frac{1}{K_w \beta_m} \right) \frac{\exp(\beta_m l_1)}{D_{1m}} \right.$$

$$- \left(\frac{1}{h_2} + \frac{1}{K_w \beta_m} \right) \frac{\exp(-\beta_m l_1)}{D_{2m}} \Big] \Big\} \exp(im\omega t). \tag{43}$$

The maintenance of the heat extraction zone at constant temperature will obviously require the variation of the flow rate of the heat exchange fluid; this variability can be visualized from the expression

$$G(t) = \dot{Q}_c(t) / \{C_w[T_{wc} - T_a(t)]\}. \tag{44}$$

c. Efficiency of Heat Retrieval at Constant Temperature. The annual average efficiency of heat extraction at constant temperature from the pond can be expressed as

$$\eta = [\int Q_c(t) \, dt]/[\int S(x = 0, t) \, dt], \tag{45}$$

where the integral is carried over a period of 1 year. Consequently, we have

$$\eta = (\alpha\tau)_{\text{eff}} S_{00} - U_L(T_{\text{wc}} - T'_{\text{a0}}). \tag{46}$$

3. Performance Prediction

Extensive numerical computations have been carried out by Sodha *et al.* (1981), Kaushika and Rao (1983), and Rao (1983) to investigate the thermal behavior of the pond system. The values of insolation $S'(x = 0, t)$ and ambient air temperature $T_a(t)$ used in these computations correspond to the observed values for the year 1974 in New Delhi. Yearly Fourier coefficients are given in Tables VA and VB.

The complex amplitudes of the harmonics of solar intensity and ambient air temperature are given by

$$S'_{0m} = a'_m \exp(-i\sigma_m), \tag{47a}$$

$$T_{am} = b'_m \exp(-i\phi_m). \tag{47b}$$

The saturated vapor pressure values for the temperature range of interest (15–40°C) were obtained from steam data tables and C_1 and C'_2 were calculated by the linear curve fitting method. The heat transfer coefficients h_0, h_1, h_2, h_5, h_e, etc., were calculated by using the considerations of free and forced convections (McAdams, 1954; Duffie and

TABLE VA

Yearly Fourier Coefficients of the Solar Intensity during the Year 1974 in New Delhi, India

m	0	1	2	3	4	5	6
a'_m (W/m^2)	218.713	51.584	22.880	8.574	8.812	6.319	1.499
σ_m (deg.)		170.694	219.316	74.584	318.063	226.234	0.000

TABLE VB

Yearly Fourier Coefficients of Ambient Air Temperature during the Year 1974 in New Delhi, India

m	0	1	2	3	4	5	6
b'_m (°C)	25.000	9.195	2.987	0.462	0.378	0.284	0.014
ϕ_m (degrees)		195.980	237.323	83.623	25.334	169.881	179.999

Beckman, 1974). The values of the parameters used in these calculations follow:

$C_1 = 222.5$ N/m² °C, $C_2' = 2137.98$ N/m²,

$C_g = 1840$ J/kg °C, $h_0 = 10.84$ W/m² °C,

$C_w = 4190.0$ J/kg, $h_1 = 56.58$ W/m² °C,

$K_g = 0.519$ W/m °C, $h_2 = 48.28$ W/m²,

$K_w = 0.569$ W/m °C, $h_3 = 78.92$ W/m² °C,

$\rho_g = 2059.0$ kg/m³, $h_e = 32.34$ W/m² °C,

$\rho_w = 1000.0$ kg/m³, $\alpha = 0.9$,

$\gamma = 0.6$, $\tau = 0.94$,

$l_1 = 0.0$ m to 0.5 m in steps of 0.1 m,

$l_2 = 0.0$ m to 5.0 m in steps of 0.25 m,

$l_3 = 0.0$ m to 5.0 m in steps of 0.25 m,
$l_T = 1.25$ m to 6.0 m,

$G = 10 \times 10^{-4}, 2.0 \times 10^{-4}, 5.0 \times 10^{-4},$ and 10×10^{-4} kg/m² s,

$T_{wc} = 40$ to 100°C in steps of 10°C,

$V = 3.0$ m/s.

The variations in maximum, minimum, and average values of the
temperature of retrieved heat flux with flow rate, the total depth of
pond, and the depth of upper convection zone are shown in Fig. 8;
corresponding variations in $\dot{Q}(t)$ are shown in Fig. 9. It is seen that the
upper convective zone is something of a liability for the solar pond and
always tends to reduce the magnitudes of $T_w(t)$ and $\dot{Q}(t)$. The boiling
point of brine at a depth of 1.5 m is envisaged to be about 106°C.
Therefore only temperatures below 106°C and corresponding $\dot{Q}(t)$ values have physical significance. The limit of the physical significance is
shown by the dotted line in Fig. 8. For various applications of the solar
pond, the desirable range of the variation of the temperature of heat
extraction zone is 40 to 100°C; this indicates that the appropriate range
of variation of flow rate is 1.0×10^{-4} to 10×10^{-4} kg/m² s.

To investigate the long-term storage characteristics of the system,
the two-week values of $T_w(t)$ and $\dot{Q}(t)$ were plotted and annual curves
were obtained. The nature of these annual curves is observed to vary
markedly with the flow rate of the heat removal fluid (water), the depth

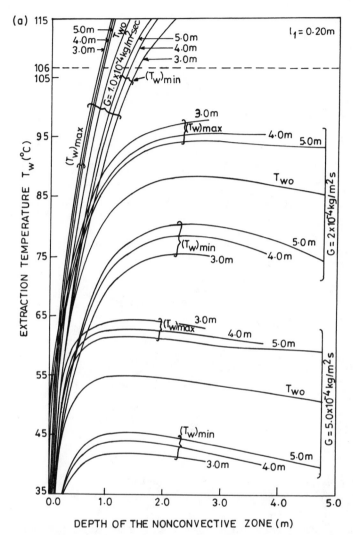

Fig. 8. (a) Variations of $(T_w)_{max}$, $(T_w)_{min}$, and T_{wo} with depth of nonconvective zone for different total depths l_T of the pond and the flow rates. The depth of the upper convective zone is kept constant (0.20 m). [After Kaushika and Rao (1983). Copyright 1984, John Wiley and Sons.] (b) Variation of T_{wo} for different depths of nonconvective zone and flow rates. [After Kaushika and Rao (1983), copyright 1984 John Wiley and Sons.]

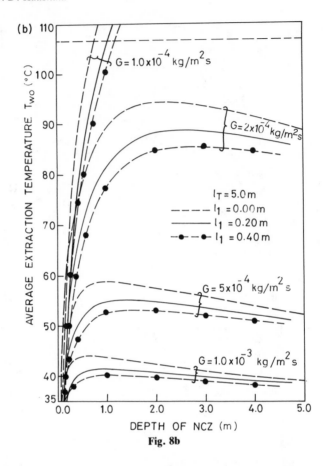

Fig. 8b

of the nonconvective zone, and the total depth of the pond. In Fig. 10, annual variations in $\dot{Q}(t)$ and $T_w(t)$ with depth of the nonconvective zone are illustrated.

Ground storage has considerable effect on the storage characteristics of the pond. To study this effect, $\dot{Q}_{max}$, Q_{min}, $(T_w)_{max}$, and $(T_w)_{min}$ have been tabulated for $K_g = 0.0$ (no ground storage) as well as for $K_g = 0.519$ W/m° C (with ground storage) in Table VI. It is observed that the presence of ground material improves the load leveling considerably.

The variation of the annual average collection efficiency at constant temperature with the depth of the nonconvective zone is illustrated in Fig. 11. With the increase in collection temperature, η decreases and the optimum value of l_2 increases. The variation of η with l_2 for differ-

TABLE VI

Effect of Ground Storage on Temperature and Heat Flux Retrieved in a Solar Pond 3 m Deep[a]

$l_T = 3.0$, $l_1 = 0.2$ m

Flow rate (kg/s m²)	Depth of nonconvective zone l_2 (m)	$(T_w)_{max}$ (°C)		$(T_w)_{min}$ (°C)		Q_{max} (W/m²)		Q_{min} (W/m²)	
		With ground	Without ground	With ground	Without ground	With ground	Without ground	With ground	Without ground
	0.00	33.13	33.27	18.17	18.02	3.92	3.75	-0.04	0.07
	0.50	77.35	79.36	56.63	54.09	41.12	41.39	29.92	28.00
2.0 × 10⁻⁴	1.00	89.85	92.16	69.26	66.02	52.28	52.97	39.81	38.10
	1.50	94.75	97.53	73.98	78.78	56.25	57.24	43.92	41.95
	2.00	96.70	100.09	75.50	70.97	57.20	58.59	46.13	44.03
	2.50	97.49	101.68	75.60	70.37	56.65	58.48	47.95	46.37
	0.00	33.12	33.25	17.95	17.81	9.34	8.92	-0.13	0.15
	0.50	60.89	62.62	38.92	36.90	64.12	64.84	42.31	40.45
5.0 × 10⁻⁴	1.00	64.17	66.20	41.86	39.41	71.40	72.60	47.99	45.54
	1.50	64.58	66.87	42.17	39.50	71.62	73.60	49.45	47.12
	2.00	64.21	66.58	41.89	39.18	69.62	72.08	50.44	49.09
	2.50	63.56	65.78	41.38	39.12	67.10	69.16	52.11	51.46
	0.00	33.09	33.22	17.64	17.51	17.31	16.52	-0.35	0.17
	0.50	50.68	51.84	28.51	27.13	79.09	81.93	47.16	46.50
1.0 × 10⁻⁴	1.00	51.38	52.68	28.80	27.35	81.93	85.42	48.99	48.33
	1.50	50.27	52.27	28.58	26.97	79.45	83.10	49.19	49.17
	2.00	50.23	51.45	28.18	26.60	75.80	78.63	50.31	51.28
	2.50	49.42	50.51	27.83	26.38	71.76	73.45	52.82	52.69

[a] Source: Rao (1983).

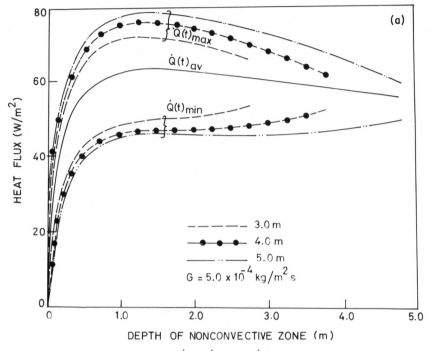

Fig. 9. (a) and (b) Variations of $\dot{Q}_{max}$, $\dot{Q}_{min}$, and $\dot{Q}(t)$ for different depths of nonconvective zone and flow rates; $l_1 = 0.2$ m. [After Kaushika and Rao (1983), copyright 1984, John Wiley and Sons.]

ent depths of the upper convective zone is illustrated in Fig. 12. The factor R corresponds to the ratio of solar heat flux at the bottom of the nonconvective zone to that incident at the pond's surface. For the solar pond without an upper convective zone ($l_2 = 0.0$ m), η_{max} is almost the same as the corresponding R; for ponds with a finite upper convective zone ($l_2 > 0.0$ m), η_{max} is lower than the corresponding value of R. An oversized pond with l_2 larger than the optimum value is more efficient than the corresponding value R. This is because at larger depths the nonconvective zone becomes a useful contributor to the collection of heat in the lower convective zone.

D. Numerical Model

It was Tybout (1967) who first pointed out the problems associated with the simplifying assumptions involved in an analytical model of a

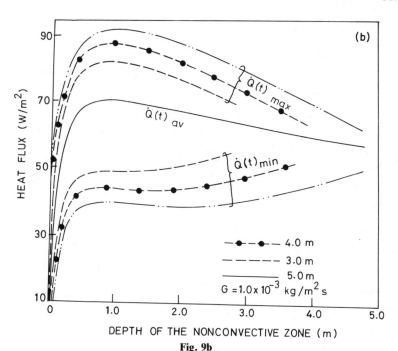

Fig. 9b

solar pond. He suggested that a numerical approach would be useful. Several authors have since employed numerical techniques and have shown that these techniques are more flexible and can handle more realistic physical situations. For example, instead of treating the insolation and ambient temperature as sinusoidal functions on a yearly cycle basis, their monthly or daily averages and even their observed hourly values could be used.

In the analytical model an expression was obtained by which the temperature and its gradient could be calculated at any point in the system. By contrast, in the numerical approach the temperature is calculated at certain discrete points of space and time. The point is regarded as representative of a certain region which includes the point. The approach amounts to trapezoidal integration of the heat conduction equation and is referred to as finite difference approximation. This is a well-known method of solving the problem of unsteady-state conduction and has been described in several publications on heat transfer (e.g., Dusinberre, 1961; Adams and Roger, 1973).

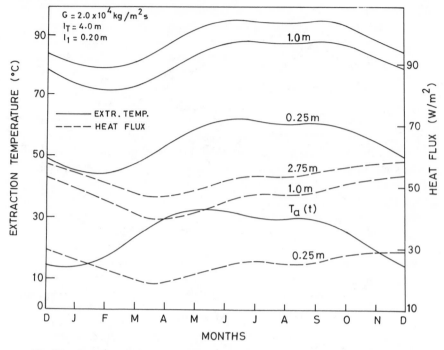

Fig. 10. Annual variations of heat flux $\dot{Q}(t)$ and the extraction temperature $T_w(t)$; $l_2 =$ 2.75 m. [After Kaushika and Rao (1983), copyright 1984, John Wiley and Sons.]

1. Finite Difference Solution

The three zones of a solar pond and a given depth of underlying ground are divided into n horizontal slices (Fig. 13); the slices need not be of equal thickness. A set of difference equations is constructed from the consideration of energy balance and thermal transfers. For kth slice the difference equation corresponding to the time interval Δt may be written as (Wang and Akbarzadeh, 1982)

$$\frac{(T_{k-1} - T_k)\,\Delta t}{\Delta x_{k-1}/2k_{k-1} + \Delta x_k/2k_k} + \frac{(T_{k+1} - T_k)\,\Delta t}{\Delta x_{k+1}/2k_{k+1} + \Delta x_k/2k_k} + \dot{q}_k\,\Delta x_k\,\Delta t$$

$$= \Delta x_k\,\rho_k C_k\,\Delta T, \tag{48}$$

where $\dot{q}$ can be expressed as $\dot{q} = -\partial S/\partial x$ in the upper convective and nonconvective zones, $\dot{q} = -(\partial S/\partial x) - \dot{Q}(t)$ in the storage zone, and $\dot{q}$

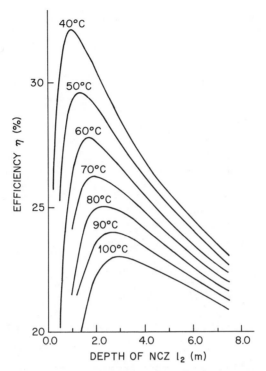

Fig. 11. Variation of thermal efficiency with collection/storage temperature and depth of nonconvective zone; $l_1 = 0.2$ m. [After Sodha *et al.* (1981), copyright 1984, John Wiley and Sons.]

$= 0$ in ground zone. If the initial temperature of each slice is known, then by solving a set of n simultaneous equations the final temperature at the end of each time interval t can be obtained and thus temperature development in the pond can be determined step by step. Thus the finite difference solution requires two boundary conditions and an initial condition. The initial condition can be determined from the physical conditions at the time of filling the pond and hence may be taken to be known. The surface boundary condition is characterized by the temperature of surface convective zone $T_1(t)$, which has been expressed in terms of climatic parameters in Eq. (18). The other boundary condition can be taken as corresponding to the constancy of groundwater temperature, which may be taken to be equal to the annual mean atmospheric air temperature.

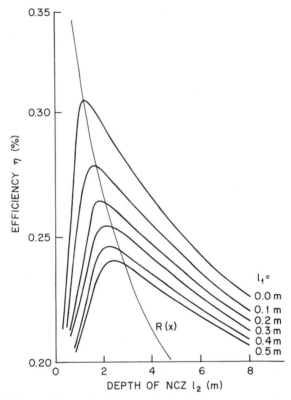

Fig. 12. Variation of thermal efficiency with depths of upper convective zone and the nonconvective zone; $T_w = 70°C$. [After Sodha *et al.* (1981), copyright 1984, John Wiley and Sons.]

2. *Results*

Hawlader and Brinkworth (1981), Sheridan (1982), Wang and Akbarzadeh (1982), and Kaushika (1983) have investigated the thermal behavior of the three-zone solar pond by using the finite difference approximation. These formulations, however, vary owing to (1) the description of $\dot{q}(x, t)$, (2) the extent of the idealization of the physical system with regard to initial and boundary conditions, and (3) the method of solution. The calculations due to Sheridan (1982) have been validated by the experimental temperature development in the solar pond at Alice Springs, Australia. He has considered the pond environment to be made up of 13 horizontal slices: (1) the upper convective

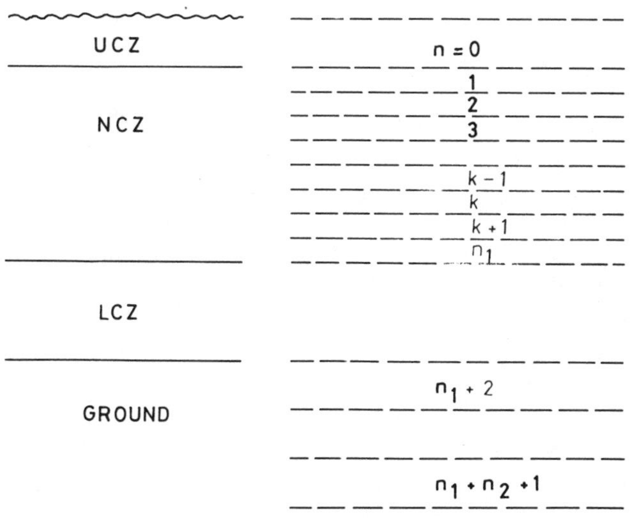

Fig. 13. Illustrations of subdivisions in solar pond for finite difference approximation.

zone is 1 slice 0.3 m thick, (2) the nonconvective zone is 1.1 m deep and made up of 11 slices, each 0.1 m thick, and (3) the lower convective zone is 1 slice 1.0 m thick. The results of the calculations are illustrated in Fig. 14; a comparison of measured and calculated temperature profiles is shown in Fig. 15.

V. CONSTRUCTION OF THE SALT-GRADIENT POND

A. Simplified Design Procedures

For sizing solar ponds, computer programs have been developed by the U.S. Solar Energy Research Institute (Edesses *et al.*, 1979) and the Victorian Solar Energy Council (Kaushika, 1982) in Australia. However, in this section a simple procedure is given for finding the depth and surface area of a solar pond for the contemplated application and location; it is intended to be a simple design tool to aid practitioners in planning and implementing solar pond energy systems.

Pond depth is made up of three zones. The surface convective zone is caused by various processes including wind mixing, penetrative convection, and diffusive action; its depth (l_1) depends on variable factors

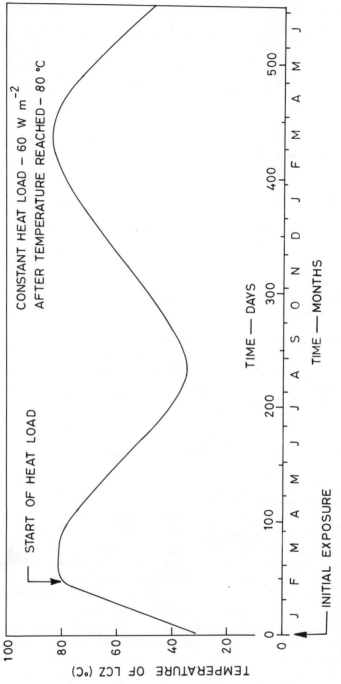

Fig. 14. Time variation of temperature of lower convective zone in solar pond at Alice Springs, Australia. [After Sheridan (1982).]

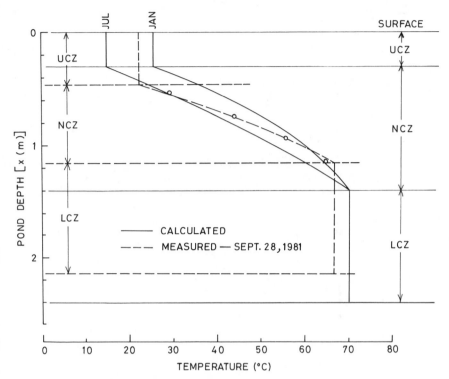

Fig. 15. Comparison of calculated and measured values of temperature in solar pond. [After Sheridan (1982).]

such as wind velocity and can grow to about 0.75 m (Atkinson and Harleman, 1983) after 1 year. However, in normal conditions a depth of 0.2 to 0.4 m is typical of practical ponds. One can design the solar pond with a l_1 of 0.4 m in mind. The bottom convective zone provides the heat storage. This zone's minimum depth is 0.5 m, a depth just sufficient to level the diurnal fluctuations in storage zone temperature. Depths above 1.5 m are very effective in leveling the seasonal variation, and for a depth of 10 m no seasonal variation in the storage zone temperature will be observed. A storage zone depth of 3.0 m gives a seasonal variation of 1.7 : 1 to 1.25 : 1. Large depths are cost intensive and a storage depth (l_3) of 1.0 to 3.0 m is, therefore, considered suitable. The nonconvective zone provides insulation. Bansal and Kaushika (1981) have investigated the insulation effect of the noncon-

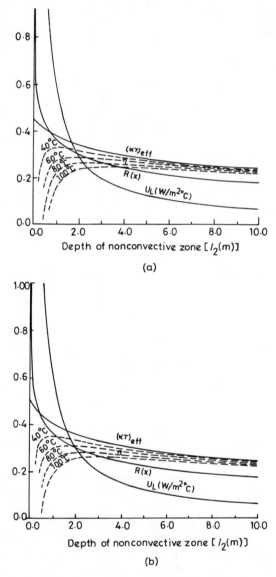

Fig. 16.   Variations of $(\alpha\tau)_{\text{eff}}$, η, $R(x)$, and U_L with the depth of nonconvective zone: (a) $l_1 = 0.4$ m; (b) $l_1 = 0.2$ m. [After Bansal and Kaushika (1981), copyright 1984, Pergamon Press Ltd.]

vective zone's depth on the heat collection efficiency and storage zone temperature. Typical results are illustrated in Fig. 16; a nonconvective zone depth (l_2) of 1.5 m seems quite appropriate to supply heat at 90 to 100°C.

The required solar pond surface area is a function of the desired annual mean pond temperature and load, the latitude of the location, and climatological parameters (viz., atmospheric air temperature and insolation). These factors influence the surface area in following manner:

1. The surface area decreases with an increase in atmospheric air temperature and insolation.
2. The surface area is larger for larger pond bottom temperatures and load requirements.
3. The surface area increases with an increase in location latitude because at higher latitudes the average incidence angle of the sun increases and so the surface reflection losses are larger; atmospheric temperature and insolation are also lower at higher latitudes. Following Edesess *et al.* (1979), the required area of a circular solar pond may be expressed as

$$A = \pi R^2 \tag{49}$$

and

$$R = \frac{U_e T_d + \{(U_e T_d)^2 + \overline{Q}_L[\overline{S}_p - (U_s + U_b)T_d]/\pi\}^{1/2}}{\overline{S}_p - (U_s + U_b)T_d}, \tag{50}$$

where $T_d = (T_{w0} - T_{a0})$, with T_{w0} the annual average pond temperature and T_{a0} the annual average atmospheric air temperature (in degrees Celsius); $\overline{Q}_L$ is the annual average load (in watts); $S_p = \tau(\alpha\tau)_{eff}\overline{S}'$; and U_s, U_e, and U_b are the average heat loss coefficients of the pond through surface, edges, and bottom, respectively.

B. Sites and Layout Considerations

The major cost factor in setting up a solar pond energy system is the construction and operation of the pond, which in turn is highly dependent upon site characteristics. Essential considerations in the selection of site are the following.

1. Water and salt are the two working materials of salt-gradient ponds. Reliable sources of water and salt are, therefore, obvious requirements at the site. Sites near salt works or concentrated salt lakes and flat, low-lying coastal lands are considered ideal. These sites are environmentally acceptable, too, since they are naturally occurring areas of salt concentration.
2. Sites subject to high winds or cyclones are to be avoided.
3. Having low-porosity soil (viz., dry soil of low thermal conductivity) with a groundwater table at a distant depth is particularly favorable since such soil reduces the brine as well as heat loss through the pond's earthen bottom and sides.
4. Arable lands are less attractive sites due to the adverse effects of land use and salt contamination on the environment.

Ponds operated to date have been built with their cross sections as circular or rectangular and their walls as vertical or slopey. The total depth of experimental ponds is 2 to 5 m, and for the structural stability of their banks and edges it is always desirable to construct ponds with slopey sides. A gradient of 45° has in general been used. Ponds on sandy soil (viz., Alice Springs in Australia) have been constructed with a low gradient of about 25° (Sheridan, 1982).

Earth removal is minimized by stacking earth to make walls around the pond. In locations where the groundwater table is very high, ponds are constructed above ground level. A vertical profile of such a pond constructed at Aspendale in Victoria (Australia) is shown in Fig. 17.

C. Containment

To prevent leakage of heat and salt from the pond, the container is made by suitable excavation and the embankment is lined with a plastic or elastometric membrane. The membrane must have good weatherability and be suitable for long service at elevated temperatures. Ethylene, propylene, diene monomer (EPDM) is the most suitable material for lining the pond. An EPDM liner (0.8 mm thick) has been successfully used in ponds built in Canada and Israel. A butynol rubber liner, supplied by Shelter Engineering Pty. Ltd. of New Zealand, is in use at the Laverton (Melbourne) solar pond. A blue liner of nylex solid vinyl (0.75 mm thick) has been used in the Alice Springs solar pond. Nylon mesh reinforced hypolon lining has also been recommended for solar ponds (Dixit, 1983).

Designing and manufacturing the membrane linings of solar ponds is

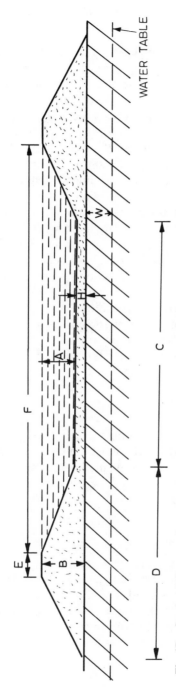

Fig. 17. Sectional diagram of solar pond at Aspendale, Victoria. $A = 0.86$ m, $B = 1.30$ m, $C = 6.63$ m, $D = 5.43$ m, $E = 0.61$ m, $F = 10.29$ m, $H = 0.3$ m, $W = 0.76$ m. [After Davey (1968).]

WATER TABLE

an expensive and difficult engineering task; the technique has not been completely satisfactory and the expensive membranes have developed leaks. Investigations to replace the membrane liners with plumbing lines and natural salt mats have, therefore, been made. The Bhavanagar (India) solar pond is lined with a high-density polyethylene membrane (0.1 mm thick) covered with clay on the bottom and brick work on the sloping walls (Kishore *et al.*, 1983). In ponds recently built in Israel the liners are made up of multiple layers of plastic packed at cores with thin layers of clay; the liner is buried under a layer of compacted earth. In many salt lakes and solar salt works a multilayered mat up to 15 cm in thickness is naturally formed. These mats are almost impervious to water and provide a natural seal against leakage (Golding, 1983) in the pond.

In some ponds a back-up liner has also been used for leakage detection and protection. The leak detection system of the Montreal (Canada) solar pond consists of sloped drainage pipes installed between the main and back-up liners; the positions of any leaks are, however, determined by a sensor grid of wires placed to monitor the electrical resistance of the sand between the liners.

Insulation can also be used in the bottom of the pond to accelerate its heating to steady-state operation.

D. Density Gradient

Various methods of establishing the salt density gradient have been discussed in Section III.C. The most convenient and expedient of all these methods is the redistribution method. Some technical data on the filling of the Miamisburg pond and the University of New Mexico pond are given in Table VII.

E. Wind Protection

Within hours of the establishment of an initial halocline (salt gradient), an erosion of the gradient into a convective layer at the surface of the pond is invariably observed. The situation is enhanced under strong wind conditions. The thickening of the surface convective zone adversely affects the thermal performance of the pond, and hence the cause and control of the formation of this top region are an important concern in solar pond study.

TABLE VII

Data on the Filling of Ponds at Miamisburg and the University of New Mexico[a]

Characteristics	UNM pond	Miamisburg pond
Surface area	175 m²	2000 m²
Depth	2.5 m	3.6 m
Total volume	230 m³	7000 m³
Diffuser size	0.3 m diameter	1.2 m diameter
Diffuser gap	3 mm	3 mm
Brine exit velocity	0.4 m/s (with small pump) / 1.5–2.6 m/s	0.6 m/s
Vertical stepping of diffuser	1–3 cm or continuous scans	5 cm
Duration of injection	0.5–4 hours per day	8–10 hours per day for seven consecutive days

[a] From Zangrando (1980). © 1980 Pergamon Press.

Wind shear generates waves and drift currents at the pond surface which tend to mix the surface water. A satisfactory theory of wind-induced mixing is yet to be developed. However, it has been observed that the dominant factor responsible for mixing the top region is the waves generated by the wind (Akbarzadeh, 1982). The wavelength of these waves depends on fetch length as well as wind velocity.

Various methods of limiting wind mixing have been tested or proposed. An outline of these methods follows.

1. Floating plastic pipes: The pond surface is covered with a floating grid of PVC pipe. This approach has been tried at Ohio State University and Bhavnagar (India) and was found to control surface waves quite effectively.
2. Floating plastic nets: This approach has been tried in recent large solar ponds in Israel.
3. Floating plastic rings: Rings made of thin plastic material and having about 0.3-m diameter and 25-mm depth have been recommended to control wind-induced waves in the Laverton solar pond (Akbarzadeh, 1982).
4. Wind baffles: Sheets of fiber glass about 0.6 m high are suspended above (by few centimeters) the water surface as shown in Fig. 18. This approach has been tried at Alice Springs.

Fig. 18.   Photograph of solar pond with wind baffles. (Courtesy Dr. R. B. Collins, Australian Solar Ponds, Alice Springs.)

All these techniques seem quite effective and economic in limiting the surface wave amplitude, but they have yet to be technically optimized. More work is, therefore, needed.

VI. MANAGEMENT OF THE SALT-GRADIENT POND

A. Monitoring

Water clarity, thermal transfer, and fluid stability are the essential aspects to be monitored in solar ponds. The physical parameters to be measured are the common meteorological parameters (temperature, insolation, humidity, and wind velocity), the temperature in the pond and the surrounding earthen region, and the solar intensity and salinity of the pond water. Temperatures are measured with the help of an array of thermocouples or thermisters generally placed at intervals of 0.1 to 0.5 m. In a simple setup, a copper–constantan thermocouple referenced to an ice-point junction can be used and the output read on a digital millivoltmeter. Sheridan (1982) has, however, reported that in the Alice Springs solar pond the mineral-insulated thermocouple with a 310 stainless sheath or with a sheath protected by a silicon skin failed due to corrosion within a few weeks. He has recommended the replacement of the silicon skin by shrinkplastic tubing.

A specific gravity bottle or picnometer can be used to measure the specific gravity of the liquid extracted from fixed levels in the pond. The radiometer may be used for the measurement of insolation in the pond. Tabor and Weinberger (1980) have suggested the use of a broad-band underwater radiometer (Kahn, 1977). This radiometer incorporates incapsulated silicon detectors and is appropriate for measurements in a hot-saline water environment.

B. Maintenance and Repair of Halocline

After filling, an important factor in pond operation is the maintenance of an appropriate salt density profile. Factors which tend to deform the profile are the vertical diffusion of salt, the surface evaporation of water, and the occurrence of localized hydrodynamic instabilities. Reconcentrating systems for gradient maintenance have been discussed in Section III.C. Evaporative losses will generate a water requirement in addition to that of the reconcentrating process. This requirement should be met with water having as low a salt content as possible. It has been suggested (Collins, 1980) that an evaporation retardation fluid, a monomolecular alcohol film, could be used to reduce the evaporation losses. The instabilities have been observed to occur in the University of New Mexico pond (Zangrando, 1979). These instabilities give rise to localized deformations in the halocline, which may be repaired by the salt "redistribution" process (Zangrando, 1979).

C. Control of Water Clarity

Water clarity is essential for good thermal performance of the pond. The factors that tend to reduce water clarity are the growth of algae or bacteria and increased levels of gilvin (Golding, 1983; Kirk, 1977) and insoluble solids such as leaves, industrial particulates, fine dust, and sand falling into the pond.

1. Chemical Treatment

Biological growth can be substantially controlled in the pond by adding a biocide (copper sulfate) to the salt in a proportion of more than 10 ppm at the time the pond is filled. Chlorination has also been found to be effective. An agent should also be used to accelerate the settling of dead organic matter.

2. *Removal of Insoluble Solids*

Insoluble solids of low specific gravity such as leaves and other organic matter float on the pond surface, and they may be removed by skimming and filtering. Sometimes breeze may move the floating material to the shore for easy removal. A filtering process may also substantially reduce the suspended fine dust particles in the surface zone.

Particulates of high specific gravity sink to the bottom and cause no harm.

VII. SHALLOW SOLAR POND

A. Configurations

The earliest experiments to demonstrate the large collection of solar energy in a shallow body of water were those of Willsie (1909) and Shuman (1909); in these experiments solar energy utilization for a water-pumping application was demonstrated. The pioneering work of Willsie and Shuman has since been substantiated by several investigations (Harris *et al.*, 1965; Gopfforth *et al.*, 1968; Khanna, 1973; Kudish and Wolf, 1978). Extensive studies of shallow solar ponds (SSPs) have since been conducted at the Lawrence Livermore Laboratory (LLL) (Day *et al.*, 1975; Dickinson and Neifert, 1975; Dickinson *et al.*, 1976; Cassamajor *et al.*, 1979; Cassamajor and Parson, 1979; Dickinson and Brown, 1979; Clark and Dickinson, 1980). More recently, some investigations of compact-type shallow solar ponds have been carried out at the Indian Institute of Technology, Delhi (Sodha *et al.*, 1980; Bansal, 1983; Kaushika *et al.*, 1982).

A shallow solar pond essentially consists of a body of convective water with a transparent cover and black bottom. The system is well insulated from the bottom and sides; surface heat exchange with the environment is reduced by the transparent cover. Water is used for three functions: (1) the collection of solar radiant energy and its conversion into heat, (2) the concentration and storage of heat, and (3) the transport of thermal energy out of the system. The high specific heat and excellent heat transfer as well as the fluid-dynamic properties of water ensure the efficient accomplishment of all the three functions. Several shallow solar pond designs are now known. The variations in these designs are essentially due to the fabrication process and material, the pond's compactness, and its application. A detailed description of design, construction, and material specifications of SSP modules tested at the Lawrence Livermore Laboratory is given in Cassamajor and Parson (1979). The schematic of a preferred module is given in Fig. 19a. It has evolved as a result of considerable experimentation. The top glazing is of corrugated fiber glass. A tight seal between

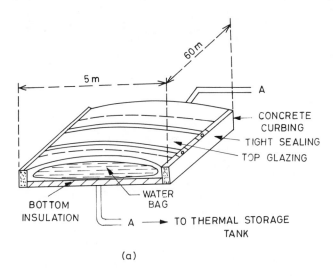

(a)

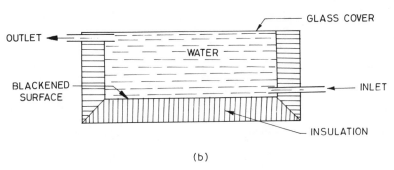

(b)

Fig. 19. Schematics of typical shallow solar pond modules tested at (a) Lawrence Livermore Laboratory (Clark and Dickinson, 1980) and (b) the Indian Institute of Technology, New Delhi (Sodha *et al.,* 1980).

the top glazing and concrete curbing is ensured by means of sheet metal brackets and conduit clamps. A schematic of the compact SSP module tested at IIT Delhi is shown in Fig. 19b. It consists of an open insulated shallow tank whose interior is blackened and whose surface is covered with glass.

B. Hot Water Withdrawal

Heat may be retrieved from the system by withdrawal of hot water. In order to have an idea of the nature of heat that can be retrieved, we consider the following two modes of hot water withdrawal.

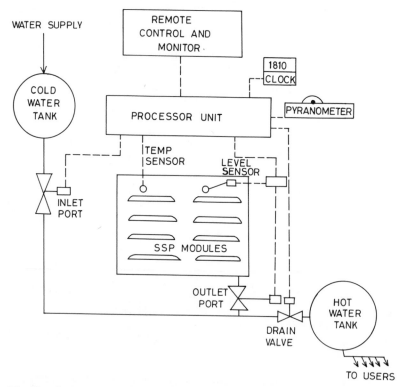

Fig. 20. System schematic for batch withdrawal operation of shallow solar pond. [After Cassamajor and Parson (1982) as quoted by Bansal (1982).]

1. *Batch Withdrawal*

The pond is filled in the morning and all of the hot water is withdrawn in the afternoon when the temperature is at its maximum. Heat is thus retrieved at constant temperature.

2. *Constant Flow Mode*

In this mode of operation hot water is withdrawn continuously at constant flow rate and the level of the water in the system is maintained constant. The temperature of the water withdrawn from the system varies with the time of the day.

Bansal (1982) has presented the schematics of control and flow arrangements (Cassamajor and Parson, 1979) for batch as well as flow operation. Figures 20 and 21 illustrate the system schematics for batch withdrawal and constant flow withdrawal operations, respectively.

C. Simple Transient Thermal Model

Incident solar radiation after passing through the top glazing is absorbed in the water mass, which gets heated. The temperature of hot water is governed by the energy balance in the system. If $\dot{m}_w$ is the flow rate at which the hot water at temperature T_o is withdrawn or the cold water at temperature T_{in} is added, then during the hours of sunshine the energy balance of the water mass in the pond module may be written as (Kaushika *et al.*, 1982)

$$\begin{bmatrix} \text{Total heat} \\ \text{taken in by} \\ \text{the absorber} \\ \text{at any time } t \end{bmatrix} = \begin{bmatrix} \text{Heat utilized} \\ \text{to increase} \\ \text{the water} \\ \text{temperature} \end{bmatrix} + \begin{bmatrix} \text{Heat lost} \\ \text{to the} \\ \text{surroundings} \end{bmatrix}$$

$$+ \begin{bmatrix} \text{Heat carried} \\ \text{away by the} \\ \text{moving fluid} \end{bmatrix}.$$

Mathematically,

$$(\alpha\tau)_{\text{eff}} \left[\int_0^t S(t) \, dt \right] A = M_w C_w (T_w - T_i) + \dot{m}_w c_w (T_o - T_i)t$$

$$+ U'_L \left(\frac{T_w + T_i}{2} - \frac{1}{t} \int_0^t T_a(t) \, dt \right) tA. \quad (51)$$

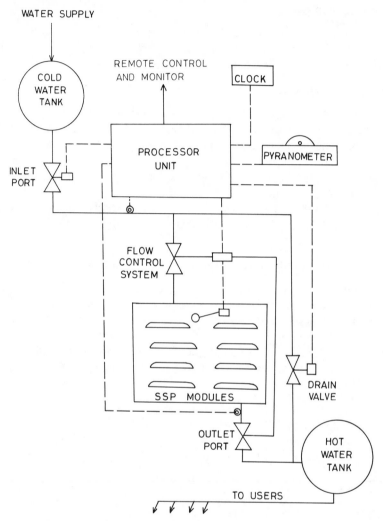

Fig. 21. System schematic for constant flow operation of shallow solar pond. [After Cassamajor and Parson (1979) as quoted by Bansal (1982).]

The change in temperature dT of the tank water during the interval dt is given by the heat balance consideration as heat gained by the flowing water equals heat lost by the hot water in the system

$$C_w(T - T_{in})\dot{m}_w \, dt = -M_w C_w \, dT,$$

which on integration yields

$$T_o = T_{in} + (T_w - T_{in}) \exp[-(\dot{m}_w/M_w)t]. \tag{52}$$

During off-sunshine hours the system is covered with insulation. If T_f is water temperature at the time of the placement of insulation, the water temperature at subsequent time intervals t' is given by the heat balance equation

$$M_w C_w(T_f - T_w) = U''_L \left(\frac{T_f + T_w}{2} - \frac{1}{t'} \int_0^{t'} T_a(t') \, dt' \right) t'A$$

$$+ \dot{m}_w c_w(T_o - T_{in})t'. \tag{53}$$

Equations (51)–(53) enable the calculation of the temperature of hot water in the system. Kaushika *et al.* (1982) have made calculations corresponding to the solar insolation and atmospheric air temperature data of April 28, 1979, at New Delhi (Fig. 22a) by using the following typical set of system parameters:

$$(\alpha\tau)_{eff} = 0.8,$$
$$U'_L = 2.5 \text{ W/m}^2 \text{ °C (for day)},$$
$$U''_L = 0.8 \text{ W/m}^2 \text{ °C (for night)},$$
$$M_w = 100 \text{ kg},$$
$$T(t = 0) = 33°C,$$
$$C_w = 4190 \text{ J/kg °C},$$
$$\dot{m}_w = \text{variable 0 to } 5 \times 10^{-2} \text{ kg/s}.$$

The results of the calculations are depicted in Fig. 22b. The curve corresponding to $\dot{m}_w = 0$ obviously provides temperature values for the case of batch withdrawal; other curves correspond to hot water withdrawal at constant flow rate.

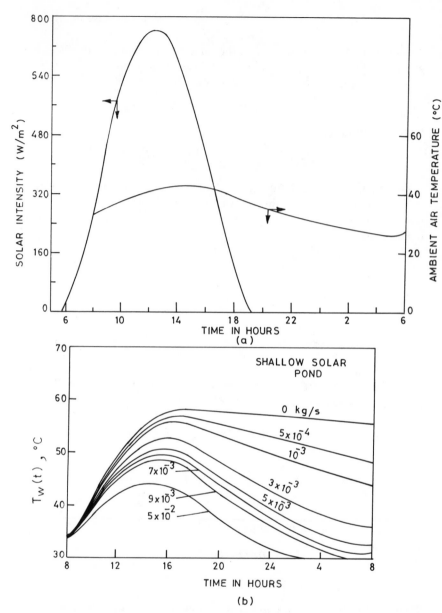

Fig. 22. (a) Hourly variation of solar intensity and atmospheric air temperature at New Delhi. (b) Corresponding variations of shallow solar pond water temperature for different flow rates of heat removal fluid. [After Kaushika *et al.* (1982), copyright 1984, Applied Science Publishers Ltd.]

VIII. ALTERNATIVE NONCONVECTIVE SOLAR PONDS

A. Operational Problems with Salt-Gradient Solar Ponds

Salt-gradient solar ponds have been successfully operated in several parts of the world and are characterized by several attractive features such as low cost per unit pond area, long-term storage (spanning seasons), and simple technology; yet there are several operational problems affecting their widespread use. Some of the problems can be outlined as below.

1. Upward diffusion of salt in the pond tends to deplete the salt gradient; the existing methods for the maintenance of salt are rather complicated and are not yet adaptable to automatic control processes. Consequently, the solar pond technology cannot be easily marketed and transferred from one region to other.
2. Nearly $\frac{1}{3}$ ton of salt per unit area is required in the construction of a salt-gradient pond and at certain locations salt contamination may cause serious environmental hazards, viz., the detrimental effect of salt in drinking water and the toxicity to vegetation surface by saline soil.
3. Wind-induced disturbances and currents stir the pond surface and form a convective zone; in preceding sections it has been shown that the surface convective zone considerably impairs the thermal efficiency of the pond.
4. High salt content combined with high temperature are corrosive to pond hardwares and, therefore, have adverse effects on the cost and life span of the pond.

In recent years several alternative solar pond concepts have been advanced to circumvent the preceding problems. This section presents a discussion on the current state of knowledge regarding the technical viability of some of these concepts, viz., the viscosity-stabilized solar pond, the membrane pond, and the honeycomb-stabilized solar pond.

B. Viscosity-Stabilized Solar Ponds

1. Concept and Approach

The idea of viscosity stabilization for a nonconvective solar pond was presented by Shaffer (1975). He suggested that a pond could be rendered nonconvective by increasing the viscosity of the pond with suitable additives and that the thermal gradients could be maintained in such a pond. This pond may better be described as static rather than stable; and hence it is seen to be free from several operational problems encountered in salt-gradient solar ponds. Physically, a viscosity-stabilized solar pond can be characterized as a horizontal layer of viscous fluid embodying a temperature gradient with maximum temperature in the bottom and minimum at the surface. The phenomenon of thermal convection in such a fluid layer has been known for about two centuries. In 1900 Bennard experimentally demonstrated that the thermal convection in the fluid layer resulted from hydrodynamic instabilities.

The temperature gradient to be maintained in a solar pond may be referred to as adverse since because of thermal expansion the fluid at bottom will be lighter than the fluid at top, constituting a potentially unstable system (Chandrasekhar, 1961). The density gradient which arises as a result of the temperature gradient provides a driving force for fluid instabilities. This tendency of the fluid is opposed by its own viscosity. It is, therefore, expected that the adverse temperature gradient must exceed a certain value before the thermal convection can set in. Rayleigh showed that the onset of instabilities is decided by the numerical value of a nondimensional quantity, called the Rayleigh number R. It relates buoyant forces and viscous drag and is expressed as

$$R = \frac{g\beta(\text{gradient})}{\kappa\nu} d^4 \quad \text{or} \quad R = \frac{g\beta(dT/dx)}{\kappa\nu} d^4,$$

and if the gradient is constant,

$$R = \frac{g\beta(\Delta T)}{\kappa\nu} d^3, \tag{54}$$

where g is acceleration due to gravity, β the coefficient of volumetric expansion of fluid, dT/dx the adverse temperature gradient, d the depth of fluid layer, κ the thermal diffusivity, and ν the kinematic coefficient of viscosity. Rayleigh has further shown that the instability sets in only when R exceeds a critical value R_c. This criterion is called the "Rayleigh criterion." Here R_c can be determined from the theory of thermohydrodynamic stability of a fluid layer with adverse temperature gradient. This problem is often referred to as the Rayleigh–Bennard problem. The mathematical treatment of this problem is available in several research articles and texts (viz., Pellow and Southwell, 1940; Chandrasekhar, 1961) wherein the accompanying values of R_c have been reported.

Number	Nature of surfaces bounding the fluid layer	R_c
1	Both rigid	1707.7
2	Bottom rigid and surface free	1100.6
3	Both free	657.5

Following Eq. (54), we have

$$(\Delta T)_{max} = R_c K\nu/g\beta d^3, \tag{55}$$

and the nonconvecting depth is

$$d = (R_c K\nu/g\beta(dT/dx))^{1/4}. \tag{56}$$

Rao (1983) has made calculations for $(\Delta T)_{max}$ corresponding to fluid layers of depths in the range 1–50 cm and kinematic viscosity in the range 10^{-7}–1.0 m^2/s (0.1–10^6 centistokes). Results of calculations corresponding to $R_c = 1700.7$ are shown in Table VIII. Similar calculations for the noncirculating depths for dT/dx in the range 0.5–2.0°C/cm have also been made, and the results are shown in Table IX. At this stage it may also be mentioned that Shaffer (1978) had shown that simple water-soluble gums can produce syrups that have kinematic viscosities in the range of 3×10^{-2} m^2/s to 30.0 m^2/s and that very low concentrations of synthetic or semisynthetic polymers can produce viscosities up to 3.0×10^{-1} m^2/s. These considerations and the results shown in Tables VIII and IX indicate that the viscous ponds of shallow depths (10–20 cm) having viscosities in the range 0.1–1.0 m^2/s can indeed maintain thermal gradients of 1 to 2°C/cm. It is desirable to use a horizontal transparent cover on the pond; the cover, besides rendering the pond surface rigid, will also reduce the water losses associated with evaporation.

2. Suitable Thickeners

The thickeners suitable for stabilizing a solar pond (1) must be soluble in water and possess high thickening efficiency, (2) must have a high degree of stability in sunlight as well as in a hot aqueous environ-

TABLE VIII

Maximum Temperatures Attainable $(\Delta T)_{max}$ in a Solution Layer Bounded at Surface as Well as at Bottom ($R_c = 1707.7$) as a Function of Viscosity for Different Depths of the Layer[a]

Depth of the layer d (m)	Kinematic viscosity ν (m²/s)							
	4.400E−07	1.000E−05	5.000E−04	5.000E−03	1.250E−02	1.000E−01	5.000E−01	1.000E+00
1.000E−02	2.213E−02	5.030E−01	2.515E+01	2.515E+02	6.287E+02	5.030E+03	2.515E+04	5.029E+04
2.500E−02	1.416E−03	3.219E−02	1.610E+00	1.610E+01	4.024E+01	3.219E+02	1.610E+03	3.219E+03
7.500E−02	5.246E−05	1.192E−03	5.961E−02	5.961E−01	1.490E+00	1.192E+01	5.961E+01	1.192E+02
1.250E−01	1.133E−05	2.575E−04	1.288E−02	1.288E−01	3.219E−01	2.575E+00	1.288E+01	2.575E+01
1.750E−01	4.130E−06	9.385E−05	4.693E−03	4.693E−02	1.173E−01	9.385E−01	4.693E+00	9.385E+00
2.250E−01	1.943E−06	4.416E−05	2.208E−03	2.208E−02	5.520E−02	4.416E−01	2.208E+00	4.415E+00
2.750E−01	1.064E−06	2.419E−05	1.209E−03	1.209E−02	3.023E−02	2.419E−01	1.209E+00	2.418E+00
3.250E−01	6.447E−07	1.465E−05	7.326E−04	7.326E−03	1.832E−02	1.465E−01	7.326E−01	1.465E+00
3.750E−01	4.197E−07	9.538E−06	4.769E−04	4.769E−03	1.192E−02	9.538E−02	4.769E−01	0.953E+00
4.000E−01	3.458E−07	7.859E−06	3.930E−04	3.930E−03	9.824E−03	7.859E−02	3.930E−01	0.785E+00
4.500E−01	2.429E−07	5.520E−06	2.760E−04	2.760E−03	6.900E−03	5.520E−02	2.760E−01	0.551E+00
5.000E−01	1.771E−07	4.024E−06	2.012E−04	2.012E−03	5.030E−03	4.024E−02	2.012E−01	0.420E+00

[a] Source: Rao (1983).

TABLE IX

Noncirculating Depths d (m) for a Solution Layer Bounded at Bottom as Well as at Surface ($R_c = 1707.7$) as a Function of Temperature Gradient for Different Viscosities[a]

ν (m²/s)	Temperature gradient dT/dx (°C/m)						
	0.500E+02	0.700E+02	0.100E+03	0.120E+03	0.150E+03	0.170E+03	0.200E+03
0.440E−06	0.459E−02	0.422E−02	0.386E−02	0.369E−02	0.349E−02	0.338E−02	0.324E−02
0.100E−05	0.563E−02	0.518E−02	0.474E−02	0.452E−02	0.428E−02	0.415E−02	0.398E−02
0.100E−04	0.100E−01	0.921E−02	0.842E−02	0.805E−02	0.761E−02	0.738E−02	0.708E−02
0.100E−03	0.178E−01	0.164E−01	0.150E−01	0.143E−01	0.135E−01	0.131E−01	0.126E−01
0.500E−03	0.266E−01	0.245E−01	0.224E−01	0.214E−01	0.202E−01	0.196E−01	0.188E−01
0.100E−02	0.317E−01	0.291E−01	0.266E−01	0.254E−01	0.241E−01	0.233E−01	0.244E−01
0.500E−02	0.474E−01	0.435E−01	0.398E−01	0.380E−01	0.360E−01	0.349E−01	0.335E−01
0.100E−01	0.563E−01	0.518E−01	0.474E−01	0.452E−01	0.428E−01	0.415E−01	0.398E−01
0.125E−01	0.595E−01	0.547E−01	0.501E−01	0.478E−01	0.452E−01	0.439E−01	0.421E−01
0.250E−01	0.708E−01	0.651E−01	0.595E−01	0.569E−01	0.538E−01	0.522E−01	0.501E−01
0.100E+00	0.100E+00	0.921E−01	0.842E−01	0.805E−01	0.761E−01	0.738E−01	0.708E−01
0.500E+00	0.150E+00	0.138E+00	0.126E+00	0.120E+00	0.144E+00	0.110E+00	0.106E+00
0.100E+00	0.178E+00	0.164E+00	0.150E+00	0.143E+00	0.135E+00	0.131E+00	0.126E+00

[a] Source: Rao (1983).

ment, (3) must have good light-transmission characteristics, and (4) must be nonpolluting, noncorrosive, and inexpensive. Shaffer (1978) and Drumheller *et al.* (1975) have presented an evaluation of possible thickeners; they have recommended sodium corboxy methyl cellulose and a commercial corboxy vinyl polymer (acrylic) as suitable thickeners. At the Indian Institute of Technology, New Delhi, some laboratory experiments on the viscosity control of convection have been conducted (Kaushika *et al.*, 1982). It has been shown that there are several water-soluble natural gums, plant extracts, and synthetic/semisynthetic polymers (viz., sodium corboxy methyl cellulose and polyvinyl alcohol) which can render a water layer (10–25 cm thick) sufficiently viscous to the nonconvective. The viscous solution produced by these thickeners is very clear and colorless and possesses almost the same light-transmission characteristics as water. The price of most of these viscous additives is quite low, so a viscosity-stabilized solar pond seems to be inexpensive, too.

C. Membrane-Stratified Solar Ponds: Honeycomb Solar Ponds

Rabl and Nielsen (1974) suggested that closely spaced transparent membranes can be used as an alternative to the salt gradient for suppressing convection in solar ponds. Membrane configuration may be an array of horizontal sheets, vertical sheets, or cellular structures placed in the top part of the pond to constitute the nonconvective zone. Obviously, such a solar pond would be environmentally attractive and nearly maintenance free.

The horizontal sheet configuration has been theoretically investigated in detail by Hull (1980a). The condition for the suppression of convection in the fluid layered between a pair of horizontal membranes heated from below has also been computed by Hull, who used the consideration of Rayleigh number as in Section VIII.B.1. Some results of calculations are illustrated in Fig. 23. It is seen that separation of membranes will be about 3 to 5 mm, and hence to minimize reflective losses at the interfaces it is necessary to match the index of refraction of the fluid to that of the membrane. Water and ethanol have been recommended as the liquids for the nonconvective zone (NCZ); Teflon® FEP film is therefore the appropriate membrane. Ethanol is preferable since it has a much lower thermal conductivity and so will need fewer membranes for the stabilization of a nonconvective zone of

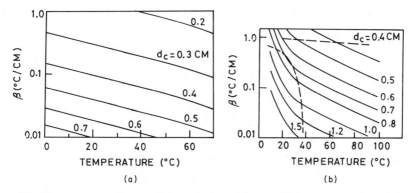

Fig. 23. Maximum distance dc between horizontal membranes as a function of temperature and temperature gradient β for (a) ethanol and (b) water. [After Hull (1980), copyright 1980, Pergamon Press Ltd.]

desired insulation. Furthermore, since ethanol is less dense than water, an NCZ with ethanol fluid can be made to float on the hot-water reservoir (the bottom convective zone) of the pond. The thermal efficiency of membrane pond has also been evaluated by Hull (1980b) and has been reported to be comparable to the salt-gradient solar pond.

Natural convection in the horizontal fluid layer heated from below may also be suppressed by the presence of vertical walls. These investigations have earlier led to the development of honeycomb structures to suppress convection in the airgap between the absorber plate and the glazing of flat-plate solar collectors (Hollands, 1965). It may, therefore, be expected that a honeycomb placed in the top region of a pond may also be an alternative device to replace the salt-gradient zone. Ortabasi *et al.* (1983) have calculated the maximum temperature gradient β_{max} that could be maintained in a honeycomb of cell size 1.25 cm and aspect ratio 8. They have reported the values of β_{max} as $4.3 \times 10^{-3}°C/m$ and $581.1°C/m$ for water and air layers, respectively. This shows that only an air-filled honeycomb would be able to maintain a temperature gradient high enough to ensure stable insulation between the hot reservoir (bottom convective zone) of the solar pond and the ambient. An air- (thermal conductivity $0.035 \text{ W/m}^2°C$) filled honeycomb zone of 10-cm thickness will provide the same thermal insulation as a nonconvective water ($K_w = 0.560 \text{ W/m}^2°C$) layer of 1.3-m thickness. On the basis of these arguments, several authors (Lin, 1982; Ortabasi *et al.*, 1983; Kaushika and Banerjee, 1983) have suggested different configurations for honeycomb-stabilized saltless solar ponds. All these configurations are shown in Fig. 24. The configuration proposed by Ortabasi *et al.* (1983) consists of open honeycomb panels (height 10–20 cm and cell size 1.2 × 1.2 cm) placed on (or partially filled with) a thin layer of silicone oil (1–5 cm) floating on the body of the hot-water reservoir. The configuration proposed by Lin (1982) comprises of a two-tier sealed, air-filled honeycomb panel floating on the body of warm water that collects solar thermal energy. The configuration due to Kaushika and Banerjee (1983) consists of the water pool contained in a closed tank made of galvanized iron or masonary walls. The tank is insulated from the sides and bottom and is blackened and glazed at the top. The honeycomb structure is placed between the glazing and blackened top surface.

The configuration due to Ortabasi *et al.* (1983) has been demonstrated to be technically viable with a laboratory model solar pond. Preliminary thermal analysis of a typical configuration with a 10-cm-high honeycomb panel and a 1-cm-thick oil layer has been carried out by Kaushika *et al.* (1983). Thermal efficiency as a function of collector parameter ($\Delta T/S_0$) has been predicted and compared with other collector and storage systems. Some results are reproduced in Fig. 25. The

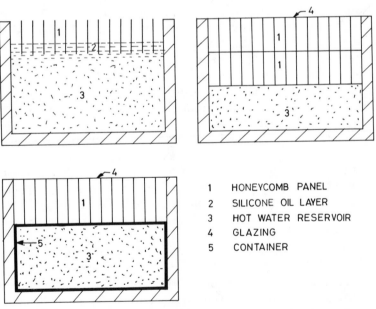

1	HONEYCOMB PANEL
2	SILICONE OIL LAYER
3	HOT WATER RESERVOIR
4	GLAZING
5	CONTAINER

Fig. 24. Proposed/tested configurations of honeycomb solar pond.

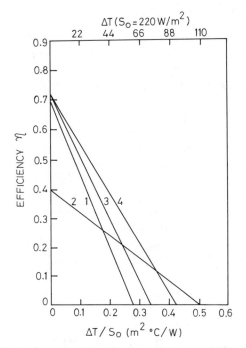

Fig. 25. The efficiency as a function of collector parameter $\Delta T/S_o$. Curve 1: flat plate collector with two glass covers and selective coating on the absorber. Curve 2: salt-gradient pond with a nonconvective zone of 1 m. Curve 3: honeycomb solar pond with side losses corresponding to side and bottom insulation as 0.075 m. Curve 4: honeycomb solar pond with side losses set equal to zero. [After Kaushika *et al.* (1983), copyright 1984, Pergamon Press Ltd.]

honeycomb solar pond seems very promising in its thermal performance.

In conclusion, the honeycomb solar pond concept seems to be technically feasible but its application to large-area ponds or lakes must certainly await further detailed study of the adverse factors, for example, the loss of honeycomb transmittance due to accumulation of dirt and other deposits with time.

IX. POSSIBLE APPLICATIONS

Heat available from salt-gradient solar ponds is at temperatures in the range of 60–100°C; the efficiency of heat retrieval from the system is about 25% at the extraction temperature 90–100°C. The solar pond can supply heat on a continuous basis, day and night, summer and winter, and so it is characterized as the only system with 100% solar fraction. Such a supply of energy holds good promise for a variety of applications, which can be divided into three classes: electric power conversion process heating, and space conditioning.

A. Electric Power Conversion

The solar pond is ideal for electricity generation application. One remarkable advantage is that its upper surface is cool due to evaporation and can be used for cooling the condenser. Thus a power generation application pond can accomplish three functions: (1) radiant energy collection, (2) thermal energy concentration and storage, and (3) heat rejection. Basic element involved in electricity generation is low-temperature turbine. Based on the nature of the turbine, two plant types are possible: (1) flashed steam plant (Fig. 26) and (2) binary fluid

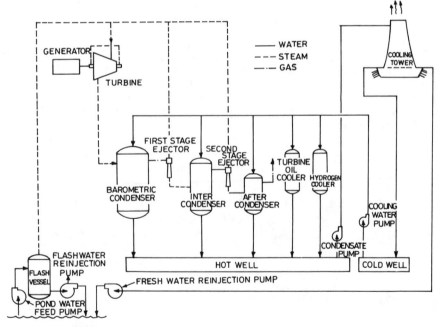

Fig. 26.   Solar pond power plant based on flashed steam cycle. [After Drumheller *et al.* (1975), copyright 1984, BNWL.]

cycle plant (Fig. 27). Commercial turbines operative with flashed steam cycle conditions are not yet available. Binary fluid cycle turbines are, however, now commercially available. In the USA, it is manufactured by the York Division of Borg-Warner Corporation for Magma Energy, Inc. (Douglas, 1975; Drumheller *et al.*, 1975). In Israel Ormat-turbines Ltd. manufactures and sells the organic Rankine cycle turbogenerators.

Organic fluid turbines are similar to steam turbines and are based on Rankine power cycle. Their working fluid is not steam but an organic fluid with low boiling temperature (such as freons), chlorinated hydrocarbons, or hydrocarbons such as toluene. These are high-molecular-weight fluids and so the turbine is smaller in size than the corresponding steam turbine. A conceptual schematic of the system is given in Fig. 28. Heat from the solar pond is supplied to an evaporator (boiler) to produce vapors that yield mechanical energy in the expander and drive an electric generator. The choice of the working substance and the type of expander is critical in the design of such a converter. Following Sheridan (1982), pressure ratios of potential working fluids are given in Table X and the dimensionless selection chart for the

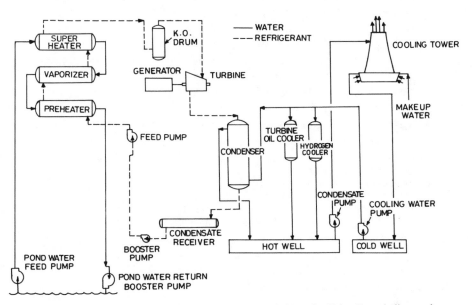

Fig. 27. Solar pond power plant based on binary fluid cycle. [After Drumheller *et al.* (1975), copyright 1984, BNWL.]

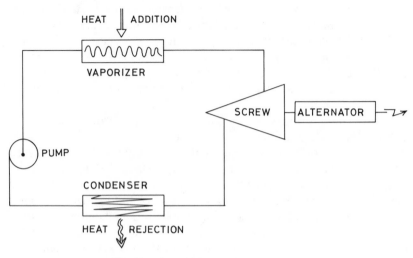

Fig. 28. Organic fluid-cycle components.

selection of expander is given in Fig. 29. Four types of expanders have been considered, viz., screw, vane, turbine, and reciprocating. It is seen that low pressure ratios are suitable with screw expander and R_{12} and R_{114} possess the lowest pressure ratio.

The first solar organic fluid turbine was operated in 1960 (Tabor and Bronicki, 1961). A 6-kW turbine was coupled to a 1500-m^2 solar pond in 1978. A 150-kW electric power plant was put in operation at Ein Bokek (Israel) in 1979. A 5-MW power module has recently been put into operation at the Dead Sea site. Efficiencies obtained with these tur-

TABLE X

Pressure Ratios of Freon Working Fluids[a]

Fluid name		Molecular weight	Vaporizer pressure (kPa)	Condenser pressure (kPa)	Pressure ratio
Commercial	Chemical				
R_{11}	CCl_3F	137.4	559	175	3.19
R_{12}	CCl_2F_2	120.9	2375	961	2.47
R_{113}	$C_2Cl_3F_3$	187.4	290	78	3.72
R_{114}	$C_2Cl_2F_4$	170.9	979	341	2.87

[a] Source: Sheridan (1982).

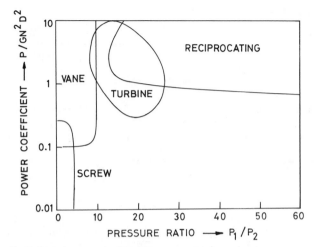

Fig. 29. Variation of power coefficient with pressure ratio. The power coefficient is a dimensionless number P/GN^2D^2, where P is power output, G is flow rate, N is speed, and D is a characteristic dimension. After Sheridan (1982).

bines are 0.4–0.7 of Carnot efficiency. A 12.5-kW solar pond electric generator has been set up at Alice Springs in Australia. This project (Sheridan, 1982) is intended to demonstrate the solar pond power application in areas remote from the main grid. The electric energy converter is a K-SPS organic rankine cycle engine based on the design of Kinetics Corporation, Florida.

The efficiency of solar to electrical conversion depends on several factors, including the transparency of the pond water, the thickness of the gradient, the nature of the underlying ground, and the turbine coupled to the pond. An estimate is possible. If we consider the utilization temperature of 90°C, the optimum efficiency for solar to thermal conversion at this temperature is 27% (Kaushika and Rao, 1983). Taking into account the practical transparencies of water and other adverse factors, a value of about 20% can be expected. Taking the sink temperature as 20°C and 7°C loss of temperature across the heat exchanger (Rao and Kaushika, 1983), the ideal Carnot efficiency is 16%. Taking the realizable power plant efficiency factor as 0.65, the efficiency of thermal to electrical conversion is obtained as about 10%. The resultant solar to electrical conversion efficiency is thus 2%; this value indicates that the solar pond is a promising development in large-scale power production technology.

TABLE XI

Solar Pond Yield Data

Yield	Pond area (m²)		
	2000	20,000	1 (km²)
Thermal energy (GW h/year)	0.657	6.57	329
Electrical energy (MW h/year)	39.4	394	19,700
Installed capacity with load factor 0.31	15 kW	150 kW	7.5 MW

Experimental efficiencies are lower than those given in the preceding. Heat experiments with a 1100-m² pond in Israel have shown that the pond can be operated at an average efficiency of 15%. For thermal to electrical conversion an efficiency of about 6% is typical. Taking the average insolation in a tropical area as 6 kW h/m² day, we have a useful available heat of 329 kW h/m² year and an electrical output of 19.7 kW h/m² year, which corresponds to 2.2 W/m² or 7.5 W/m² with a load factor of 0.3.

Some useful solar pond yield data are tabulated in Table XI.

B. Process Heating

Edesess (1982) has made a worldwide survey of sites and applications for salt-gradient solar ponds. Several sites have been examined. These include salinas, salt flats and salt marshes including Sebkhas Moknine in Tunishia; terminal lakes such as the Dead Sea; brine impoundment ponds; industrial and mining impoundment ponds, and site-built solar ponds. It has been shown that these ponds in general and the site-built ponds in particular hold great promise for process heating applications. Various sectors may now be identified for solar pond application.

1. Industrial Sector

Gerofi and Fenton (1981) have surveyed the general potential for the utilization of solar pond energy in the Australian industrial sector and have pointed out that under present economic conditions there are locations at which heat energy from a solar pond is considerably cheaper than that from fossil fuels. Any of the following industries and industrial processes could be supplied heat from a solar pond: salt and mineral production; drying and curing of timber; milk pasteurization; concentration and separation by evaporation; cleaning and washing in the food industry; tanning of skins; textile processing such as wool scouring, carbonizing, and dyeing; industrial laundry; paper industry for preheating.

Salt production is one of the earliest applications of the solar pond (Davey, 1968; Hirschmann, 1970; Matz et al. 1965). A simplified diagram of the solar pond process for salt making is given in Fig. 30.

Mitre Corporation (1973) has reported that in the United States about two-thirds of the total industrial energy goes into process heating; food and paper industries consume about 10% of it. The paper industry uses thermal energy at a temperature of about 200°C. Therefore solar pond energy could be used for preheating (Styris *et al.*, 1975). A simple schematic of the arrangement to accomplish the task is given in Fig. 31.

2. Agricultural Sector

The solar pond heat has multiple applications in the agricultural industry. However, extra care must be taken to avert the danger of salt contamination of surface soil as well as underground water. Some of the recommended uses are crop drying on farms, heating stock animal housing, heating the greenhouse environment, providing heat for boosting the performance of biogas plants and providing heat for biomass conversion to alcohol fuels etc.

At the Ohio Agricultural Research and Development center (OARDC) a solar pond has been set up and operated to control the thermal environment of a greenhouse (Badger *et al.*, 1977; Short *et al.*, 1978). A photograph of a solar pond at the OARDC in Wooster, Ohio, is given in Fig. 32. The pond is used to heat the greenhouse behind it; this view is taken from south, that is, looking due north.

3. Social Sector

Solar pond heat has been demonstrated to be applicable to the social sectors of both the developed and the less-developed world. Specific applications include the following: heating swimming pools, low-temperature cooking in southern India and western Africa, desalinating water, viz., in outback Australia.

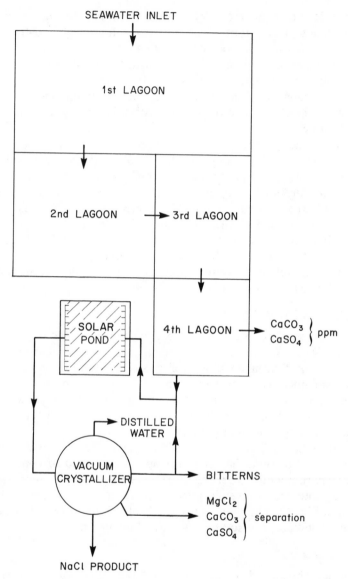

Fig. 30.   Schematic of solar pond process for salt production. [After Davey (1968).]

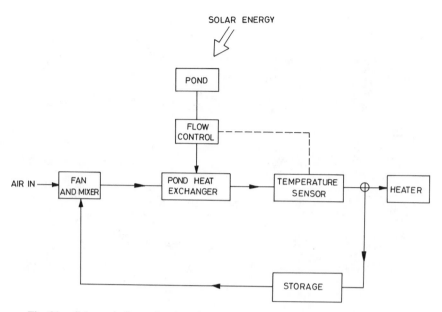

Fig. 31. Schematic for preheating air for process heat application in paper industry. [After Styris *et al.* (1975), copyright 1984, BNWL.]

The 2000-m^2 Miamisburg, Ohio, solar pond has been constructed and operated (Bryant *et al.*, 1979; Wittenberg and Harris, 1981) to heat a swimming pool and a recreational building. This solar pond costs about U.S. $84,000, and its performance paid back its price in 15 years.

C. Space Conditioning

The application of solar pond heat in space conditioning is an example of system application penetration to urban areas. It could be economically viable only if land prices are not prohibitive. The central heating of buildings (Rabl and Nielsen, 1975; Styris *et al.*, 1975, 1976) is one mode of application. The other could be the utilization of the solar pond as a constant heat source for a solar absorption space conditioning system (viz., Grassie and Sheridan, 1977) as well as for Rankine cycle solar engines (Teagan, 1973); Tabor and Weinberger (1980) have indicated that it is a viable and promising application of solar pond heat.

Fig. 32.   Photograph of solar pond at OARDC, Wooster, Ohio. The pond is used to heat the greenhouse behind it. The agricultural engineering building is behind the greenhouse. The city of Wooster is on the hill in the background. (Courtesy Dr. Peter Fynn, OARDC.)

X. ECONOMICS

A meaningful economic analysis of the solar pond requires worldwide operational and maintenance data which are available only for salt-gradient solar (SGS) ponds. Most of SGS ponds constructed to date have been prototype devices meant for developing the technology and identifying the problems. The actual cost of such constructions is not of great significance except insofar as they are used as an indicator of the likely costs that might be achieved by experienced builders.

The major cost factors of a solar pond system are the (1) the earth work (excavation, leveling, and compaction), (2) the synthetic lining, (3) the salt, (4) the water, (5) the land, and (6) other miscellaneous supplies. These factors considerably vary with the location as well as the size of the system. Consequently, small ponds are considerably more expensive than large ponds and a wide variability is expected in

TABLE XII

Costs of Small-Size Solar Pond Installations

Specification	Australian pond[a]	American pond[b]
Earth work (U.S. $/m²)	9.81	5.00
Liner (U.S. $/m²)	2.79	11.00
Salt (U.S. $/m²)	1.91	9.70
Water (U.S. $/m²)	0.025	—
Land	—	—
Miscellaneous (U.S. $/m²)	0.52	5.90
Total (U.S. $/m²)	15.06	31.60

[a] Location: Alice Springs (Sheridan, 1982).
[b] Location: Miamisburg (Wittenberg and Harris, 1981).

the per unit area cost estimates of solar ponds operated at different locations around the world. Installation costs of two small-size solar ponds are illustrated in Table XII. These estimates exclude all expenses entailed with the heat recovery systems.

According to Tabor (1981), in 1979 the SOLMAT Company calculated that a salt-gradient solar pond could be built in most areas for about U.S. $13/m² with a liner and for U.S. $8/m² without a liner. These figures correspond to ponds of area about 1×10^5 m².

The cost of energy from solar ponds depends upon the annual mean insolation, the recurring expenditure on maintenance, and the annual charges on capital. Edesess (1982) has considered three typical cases: best case, conservative estimate, and worst case. The conservative estimate is characterized by the most detailed up to date site-specific design studies. Best case is the upper boundary of the realistic possibility, and the worst case is a pessimistic version of conservative estimate. Assuming that the annual charge on capital is 15%, Edesess (1982) has reported the cost of electricity as 5.1¢, 20.5¢, and 81.5¢/kW h for the best, conservative, and worst cases, respectively. The estimated cost of process heat is 0.59, 4.78, and 10.71 U.S. $/gJ for the three cases, respectively. With the same (15%) cost, Sheridan (1982) has estimated the electrical energy cost for a 15-kW pond as 60.0¢/kW h and the thermal energy cost as 2.4 U.S. $/gJ.

The preceding costs of energies should be compared with alternatives. According to Multer (1980), in 1980 the cost of thermal energy from fuel oil was about U.S. $11.37/gJ. The cost of coal heat was about the same and that of electric heat was more than oil. This indicates a high potential for solar pond heat utilization. Solar pond electrical energy costs seem large compared to cost on the electricity grid (U.S. $0.06/kW h). However, in Australia a cost of about $1.00/kW h is considered typical for remote areas (Sheridan, 1982). Thus pond electricity costs seem to be of the right order for remote areas.

NOMENCLATURE

A	Pond area (m^2)
C_1	Slope of saturated vapor pressure and temperature curve (N/m^2 °C)
C_2'	Intercept of saturated vapor pressure and temperature curve (N/m^2)
C_g	Specific heat of ground (J/kg °C)
C_w	Specific heat of water (J/kg °C)
G	Fluid flow rate per unit pond area (kg/s m^2)
h_0	Radiative plus convective heat transfer coefficient from the pond surface to the ambient (W/m^2 °C)
h_1	Heat transfer coefficient between the upper convective zone and the nonconvective zone water of the pond system (W/m^2 °C)
h_2	Heat transfer coefficient between the nonconvective zone and the lower convective zone water of the pond system (W/m^2 °C)
h_3	Heat transfer coefficient between the lower convective zone water and the blackened surface (ground zone) of the pond system (W/m^2 °C)
ha	Hectare
h_e	Heat transfer coefficient of evaporation at the pond surface (W/N)
k	Variable number of slices in finite difference approximation
K_w	Thermal conductivity of water (W/m °C)
K_g	Thermal conductivity of ground (W/m °C)
l_1	Depth of the upper convective zone (m)
l_2	Depth of the nonconvective zone (m)
l_3	Depth of the lower convective zone (m)
m	Number of harmonics
M_w	Mass of water in shallow solar pond (kg)

$\dot{m}_w$	Heat removal fluid flow rate (kg/s)
n	Absorption coefficient for the entire spectrum of solar radiation (m^{-1})
n_j	Absorption coefficient for the jth portion of the solar spectrum (m^{-1})
P_r	Prandtl number
$\dot{Q}(t)$	Retrieved heat flux from the lower convective zone of the solar pond (W/m^2)
$\dot{Q}_c(t)$	Retrieved heat flux at constant temperature of the heat extraction zone (W/m^2)
$\dot{Q}_e$	Heat loss due to evaporation (W/m^2)
$\dot{Q}_L$	Annual average load (W)
r	Effective angle of refraction at the pond surface
S	Salinity (kg/m^3)
$S(x, t)$	Solar intensity at a point specified by x at time t (W/m^2)
$S(x = 0, t)$	Solar intensity at the pond surface at time t (W/m^2)
S_{0m}	$\tau S'_{0m}$ (W/m^2)
S'_{0m}	Amplitude of the mth harmonic of solar intensity at the pond surface (W/m^2)
S_0	τS_{00} (W/m^2)
S_{00}	Average value of solar intensity $S(x = 0, t)$ (W/m^2)
$T(x, t)$	Temperature distribution in the pond system at a point specified by x at time t (°C)
$T_a(t)$	Ambient air temperature (°C)
T_{a0}	Average value of $T_a(t)$ (°C)
T_{am}	Amplitude of the mth harmonic of $T_a(t)$ (°C)
$T_i(t)$	Inlet water temperature (°C)
$T_o(t)$	Outlet water temperature (°C)
T_{x0}	Average value of $T(x, t)$ (°C)
$T_w(t)$	Water temperature in the bottom convective zone of the pond system (°C)
T_{w0}	Average value of $T_w(t)$ (°C)
T_{wm}	Amplitude of the mth harmonic of $T_w(t)$ (°C)
$T_1(t)$	Temperature of the upper convective zone (°C)
t	Time coordinate (s)
U'_L	Heat loss coefficient of shallow solar pond during the day (W/m^2 °C)

U_L''	Heat loss coefficient of shallow solar pond during the night (W/m² °C)
x	Position coordinate (vertically downward) (m)
Z_w	Heat capacity of lower convective zone brine per unit pond volume (J/m³ °C)
α	Absorptance of the blackened surface
α_S	Coefficient of salt diffusion (m²/s)
α_T	Coefficient of temperature diffusion
α_c	Thermal expansion coefficient
β_c	Salt expansion coefficient
η	Annual average efficiency of heat retrieval
γ	Relative humidity
μ_j	Fraction of solar radiation having absorption coefficient n_j
ν	Kinematic viscosity of pond water (m²/s)
ρ_w	Density of pond water (kg/m³)
ρ_g	Density of ground (kg/m³)
τ	Coefficient of radiation transmission at the pond surface
τ_s	Ratio of salt diffusivity to temperature diffusivity (α_S/α_T)

Abbreviations

ASP	Alternative solar ponds	SSP	Shallow solar pond
LCZ	Lower convective zone	SGSP	Salt-gradient solar pond
NCZ	Nonconvective zone	UCZ	Upper convective zone

ACKNOWLEDGMENTS

The author sincerely thanks Drs. N. R. Sheridan, W. W. S. Charters, S. C. Kavshik, S. K. Rao, and P. K. Bansal and Mr. M. S. Sharma for help and suggestions. Thanks are also due to Drs. Peter Fynn and Robert Collins for sending the photographs of their experimental solar ponds.

REFERENCES

Adams, J. A., and Rogers, D. F. (1973). "Computer Aided Heat Transfer Analysis," pp. 175–227. McGraw-Hill, New York.
Akbarzadeh, A. (1982). An introduction to the design and potential of solar ponds. *Proc. ISES-ANZ*, pp. 1–6. Victorian Branch, Melbourne.
Akbarzadeh, A., and Ahmadi, G. (1980). Computer simulation of the performance of solar pond in southern part of Iran. *Sol. Energ.* **24**, 143.

Akbarzadeh, A., and Macdonald, R. W. G. (1982). Introduction of a passive method for salt replenishment in operation of solar ponds. *Sol. Energ.* **29**, 71.

Anderson, G. C. (1958). Some limnological features of a shallow saline meromictic lake. *Limnol. Oceanogr.* **3**, 259.

Assaf, G. (1976). The Dead Sea, a scheme for a solar lake. *Sol. Energ.* **18**, 294.

Atkinson, J. F., and Harleman, D. R. F. (1983). A wind-mixed layer model of solar ponds. *Sol. Energ.* **31**(3), 243.

Badger, P. C., Short, T. H., Roller, W. L., and Elwell, D. L. (1977). A prototype solar pond for heating greenhouses and rural residences. *ISES—American Section Meeting*, Orlando, Florida.

Bansal, P. K. (1983). Solar collector-cum-storage systems, *In* Reviews of Renewable Energy Resources (M. S. Sodha, S. S. Mathur, and M. A. S. Malik, eds.), Vol. I, pp. 80–134. Wiley, New York.

Bansal, P. K., and Kaushika, N. D. (1981). Salt gradient stabilized solar pond collector. *Energ. Convers. Manage.* **21**, 81–95.

Bronicki, L. Y. (1981). Low temperature turbines. *Sunworld* **5**(4), 121.

Bronicki, L. Y., Lev-Er, and Porat, Y. (1980). "Large Solar Electric Power Plant Based on Solar Ponds." World Power Conf., Munich, Federal Republic of Germany.

Bronicki, L. Y., Doron, B., Raviv, A., and Tabor, H. (1983). Progress in solar ponds. *Sol. World Congr. Perth, Ext. Abstr.* p. 168.

Brown, K. C. Edesess, M., and Jayadev, T. S. (1979). Solar ponds for industrial process heat. *Proc. Solar Industrial Process,* SERI/TP-351-460. Heat Conference at Oakland Hyatt House, Oakland, California.

Bryant, H. C., and Colbeck, I. (1977). A solar pond for London. *Sol. Energ.* **19**, 321.

Bryant, R. S. (1980). *Campus News University of New Mexico,* July 17, 1980.

Bryant, R. S., Bowser, R. P., and Wittenberg, L. J. (1979). Construction and initial operation of the Miamisburg salt-gradient solar pond. *ISEC,* Atlanta, May 28.

Carrier, W. H. (1980). The temperature of evaporation. *Trans. Am. Soc. Heat. Ventilating Eng.* **22**, 24.

Cassamajor, A. B., and Parsons, R. E. (1979). "Design Guide for Shallow Solar Ponds," UCRL-523825, Rev. 1, January 8, Lawrence Livermore Laboratory.

Cassamajor, A. B., Clark, A. F., and Parsons, R. E. (1979). "Cost Reductions and Performance Improvements for Shallow Solar Ponds," UCRL-79280, Lawrence Livermore Laboratory.

Chandrasekhar, S. (1961). "Hydrodynamic and Hydromagnetic Stability," Chapter II. Oxford University Press (Clarendon), London, and New York.

Chepurniy, V., and Savage, S. B. (1974). An analytical and experimental investigation of a laboratory solar pond model. ASME publication 74-WA/SOL-3.

Chepurniy, N., and Savage, S. B. (1975). The effect of diffusion on concentration profiles in a solar pond. *Sol. Energ.* **17**, 203.

Clark, A. F., and Dickinson, W. C. (1980). Shallow solar ponds, *In* "Solar Energy Technology Handbook" (W. C. Dickinson and P. N. Cheremisinoff, eds.), Chapter 12, p. 377.

Cohen, Y., Krumbein, W., and Shilo, M. (1977). Solar lake (Sinai). *Limnol. Oceanog.* **22**, 609.

Collins, R. B. (1981). Alice springs solar pond project. *Proc. Sol. Energ. Outback Conf.* Alice Springs, Northern Territory, Australia.

Collins, R. B. (1983). Alice springs solar pond project, *Sol. World Congr. Perth, Ext. Abstr.* p. 169.

Dake, J. M. K. (1973). The solar pond: Analytical and laboratory studies. *UNESCO Congr. Sun Service of Mankind,* Paris.

Dake, J. M. K., and Harleman, D. R. F. (1966). "An Analytical and Experimental Investigation of Thermal Stratification in Lakes and Ponds." Report No. 99, Hydrodynamics Laboratory, Massachusetts Institute of Technology.

Daniels, D. G., and Merriom, M. F. (1975). Fluid dynamics of selective withdrawal in solar ponds. *ISES Congr.,* Los Angeles, California.

Davey, T. R. A. (1968). The Aspendale solar pond. Report R15, CSIRO, Australia.

Day, J. A., Dickinson, W. C., and Iantuono, A. (1975). "Industrial Process Heat from Solar Energy." UCRL-76390, Lawrence Livermore Laboratory.

Defant, A. (1961). "Physical Oceanography," Vol. I, p. 53. Pergamon, Oxford.

Dickinson, W. C., and Brown, K. (1979). "Economic Analysis of Solar Industrial Process Heat Systems." UCRL-52814, Lawrence Livermore Laboratory.

Dickinson, W. C., and Neifert, R. D. (1975). "Parametric Performance and Cost Analysis of the Proposed S. Ohio Solar Process Heat Facility." UCRL-51783, Lawrence Livermore Laboratory.

Dickinson, W. C., Clark, A. F., Day, J. A., and Wouters, L. F. (1976). The shallow solar pond energy conversion system. *Sol. Energ.* **18,** 3–10.

Dixit, D. K. (1983). Solar ponds: Suntraps under water. *Science Today,* May, p. 31.

Douglass, L. H. (1975). Ocean thermal energy: An engineering evaluation. *Proc. Offshore Technol. Conf.,* Paper OTC 2252, Houston, Texas.

Drumheller, K., Duffy, J. B., Harling, O. K., Knutsen, C. A., McKinnon, M. A., Peterson, P. L., Shaffer, L. H., Styris, D. L., and Zaworski, R. (1975). "Comparison of Solar Pond Concepts for Electrical Power Generation." BNWL-1951, Battelle-Northwest Laboratories, Richland, Washington.

Duffie, J. A., and Beckman, W. A. (1974). "Solar Energy Thermal Processes." Wiley, New York.

Dusinberre, M. (1961). "Heat Transfer Calculations by Finite Difference." International Text Book Company.

Edesess, M. (1982). On solar ponds: Salty fare for the world's energy appetite. *Tech. Rev.*

Edesess, M., Henderson, J., and Jayadev, T. S. (1979). "A Simple Design Tool for Sizing Solar Ponds." SERI/RR-351-347, Golden, Colorado.

Elata, C., and Levin, O. (1962). "Selective Flow in a Pond with Density Gradient." Hydraulic Laboratory Rep., Technion, Haifa, Israel.

Elata, C., and Levin O. (1965). Hydraulics of the solar pond. *11th Congr. Int. Assoc. Hydraul. Res.,* Leningrad, USSR.

Eliseev, V. N., Usmanov, Yu. U., and Teslenko, L. N. (1971). Theoretical investigation of the thermal regime of a solar pond. *Geliotekhnika* **7**(4), 17.

Eliseev, V. N., Usmanov, Yu. U., and Umarov, Ya. (1973). Determining the efficiency of a solar pond. *Geliotekhnika* **9**(1), 44.

Elwell, D. L., Short, T. H., and Badger, P. C. (1977). Stability criteria for solar (thermal saline) ponds. *Proc. Annual Meeting American Section of ISES,* Orlando, Florida.

Gerofi, J. P., and Fenton, G. G. (1982). "Solar Ponds Current Status and Potential in N.S.W." Report by ENERSOL Consulting Engineers to Energy Authority of New South Wales.

Golding, P. (1983). Meromictic water reservoir solar collectors. *Sol. World Congr. Perth, Ext. Abstr.,* p. 229.

Golding, P., Akbarzadeh, A., Davey, J. A., MacDonald, R. W. G., and Charters, W. W. S. (1982). Design features and construction of Laverton solar ponds. *ISES ANZ Sect. Conf.* "Solar Energy Coming of Age." Brisbane, pp. 33–36.

Golding, P., Davey, J. A., Macdonald, R. W. G., Charters, W. W. S., and Akbarzadeh, A. (1983). Construction and operation of Laverton solar ponds. *Sol. World Congr. Perth, Ext. Abstr.,* p. 286.

Gopfforth, W. H., Davidson, R. L., Harris, W. B., and Baird, M. J. (1968). Performance correlation of horizontal plastic solar water heaters. *Sol. Energ.* **12,** 383.

Grassie, S. L., and Sheridan, N. R. (1977). Modelling of a solar operated absorption air conditioner system with refrigerant storage. *Sol. Energ.* **19,** 691.

Gupta, C. L., and Satish, Patel (1979). Experimental investigations on laboratory solar ponds. Paper 23, *NSEC,* India.

Harris, M. J., and Wittenberg, L. J. (1979). Heat extraction from a large salt gradient solar pond. *2nd Annu. Sol. Heat. Cool. Conf.,* Colorado Springs, Colorado.

Harris, W. R., Davidson, R. R., and Hood, D. W. (1965). An experimental solar water heater. *Sol. Energ.* **9,** 193.

Hawlader, M. N. A. (1980). The influence of extinction coefficients on the effectiveness of solar ponds. *Sol. Energ.* **25,** 461.

Hawlader, M. N. A., and Brinkworth, B. J. (1981). An analysis of nonconvecting solar pond. *Sol. Energ.* **27,** 195.

Hipsher, M. S., and Boehm, R. F. (1976). Heat transfer considerations of a nonconvecting solar pond exchange. *ASME-76-WA/Sol,* 4.

Hirschmann, J. (1965). Project of a solar desalination plant for the North of Chile. *Proc. 1st Symp. Water Desalination,* Washington, D.C.

Hirschmann, J. (1970). Salt flats as solar heat collectors for industrial purposes. *Sol. Energ.* **13,** 83–97.

Hoare, R. A. (1966). Problems of heat transfer in Lake Vanda. *Nature* **210,** 787.

Hollands, K. G. T. (1965). Honeycomb devices in flat plate solar collectors. *Sol. Energ.* **9,** 159.

Huber, W. C., and Harleman, D. R. F. (1968). Laboratory and analytical studies of thermal stratification of reservoirs. Report 112, Hydrodynamics Laboratory, Massachusetts Institute of Technology.

Hudec, P. P., and Sonnefeld, P. (1974). Hot brines on Los Roques, Venezuela. *Science* **185,** 440.

Hull, J. R. (1979). "Physics of the Solar Pond. Ph.D. Thesis, Iowa State University, Ames, Iowa.

Hull, J. R. (1980a). Membrane stratified solar ponds. *Sol. Energ.* **25,** 317.

Hull, J. R. (1980b). Computer simulation of solar pond thermal behaviour. *Sol. Energ.* **25,** 33.

Hull, J. R. (1982). Calculation of solar pond thermal efficiency with a diffusely reflecting bottom. *Sol. Energ.* **29**(5), 385.

Jain, G. C. (1973). Heating of solar ponds. *UNESCO Congr. Sun Service of Mankind,* Paris.

Jayadev, T. S., and Edesess, M. (1980). "Solar Ponds." SERI/TP-731-587, Golden, Colorado.

Jayadev, T. S., Edesess, M., and Henderson, J. (1979). Solar Pond Concepts: Old and New. *14th Int. Soc. Energy Convers.* SERI/TP-35-213, Eng. Conference, Boston, Massachusetts, Aug. 1979.

Kahn, M. (1977). "Broadband Underwater Radiometer." B. S. Thesis, Jerusalem College of Technology, Jerusalem, Israel.

Kalecsinsky, A. V. (1902). Ueber die ungarischen warmen and heissen Kochsalzseenals natuerliche Waermeaccumulatoren. *Ann. Phys.* **IV**(7), 408–416.

Karnal, J., and Nielsen, C. E. (1983). Zone boundary fluctuations in solar ponds. *Sol. World Congr. Perth Ext. Abstr.*, p. 245.

Katti, Y., and Kaushika, N. D. (1981). Thermohydrodynamic instabilities in salt gradient solar ponds. *Proc. N.S.E.C.*, Bangalore, India.

Kaushika, N. D. (1982). Thermal design and performance prediction of salt gradient solar pond. *Rep. Victorian Sol. Energ. Counc.*, Melbourne, Australia.

Kaushika, N. D., and Banerjee, M. B. (1983). Honeycomb solar pond: Evaluation of applications. *Sol. World Congr. Perth, Ext. Abstr.*, p. 246.

Kaushika, N. D., and Bansal, P. K. (1982). Transient behaviour of salt gradient stabilized shallow solar ponds. *Appl. Energy* **10**(1), 47–63.

Kaushika, N. D., and Rao, S. K. (1983). Constant flow three zone solar pond collector/storage system. *Energy Res.* **7**, 61–79.

Kaushika, N. D., Bansal, P. K., and Sodha, M. S. (1980). Partitioned solar pond collector/storage system. *Appl. Energy* **7**, 169–190.

Kaushika, N. D., Bansal, P. K., and Kaushik, S. C. (1982a). Transient behaviour of collector/storage solar water heaters for generalized demand patterns. *Appl. Energy* **12**, 259–267.

Kaushika, N. D., Jaivir, Singh, Rao, S. K., and Kaushik, S. C. (1982b). Experimental observations on a viscosity stabilized solar pond. *Proc. N.S.E.C.*, New Delhi, p. 5.018.

Kaushika, N. D., Banerjee, M. B., and Katti, Y. (1983).Honeycomb solar pond collector and storage system. *Energy* **8**(11), 883.

Khanna, M. L. (1973). A portable type solar water heater. *Sol. Energ.* **15**, 269.

Kirk, J. I. (1977). Attenuation of light in natural waters. *Aust. J. Mar. Freshwater Res.* **28**, 497–508.

Kishore, V. V. N., Nielsen, C. E., Rao, K. S., and Gupta, C. L. (1983). "Rept. Programme for Solar Pond Development in India."

Kooi, C. F. (1979). The steady state salt gradient solar pond. *Sol. Energ.* **23**, 37.

Kooi, C. F. (1981). Salt gradient solar pond with reflective bottom: Application to the saturated pond. *Sol. Energ.* **26**(2), 113.

Kudish, A. I., and Wolf, D. (1978). A compact shallow solar pond hot water heater. *Sol. Energ.* **21**, 317.

Leshuk, J. P., Styris, D. L., and Harling, O. K. (1978). Solar Pond stability experiments. *Sol. Energ.* **21**, 237.

Lin, E. I. H. (1982). A saltless solar pond. *In Proc. ISES—American Section Conf. Progress in Solar Energy*, pp. 215–220.

Linke, W. F., (1965). "Solubility of Inorganic and Metal Organic Compounds," Vol. II. American Chemical Society.

Matz, R., Fiest, E., and Bloch, M. R. (1965). Production of salt by means of a solar pond. *Inst. Chem. Eng. (London) Chem. Eng. Prog.* **CE-87**.

McAdam, W. H. (1954). "Heat Transmission." McGraw-Hill, New York.

Melack, J. M., and Kilham, P. (1972). Lake Mahega: A mesotropic sulfactochloride lake in western Uganda. *Afr. J. Trop. Hydrobiol. Fish.* **2**, 141.

Mitre Corporation (1973). "Solar Energy Research Program Alternatives." No. PB-231-141, National Technical Information Service, Springfield, Virginia.

Multer, R. K. (1980). Solar pond energy systems. *ASHRAE J.* November.

Nielsen, C. E. (1975). Salt gradient solar ponds for solar energy utilization. *Environ. Conserv.* **2**, 289.

Nielsen, C. E. (1976). Experience with a prototype solar pond for space heating. *Proc. ISES Meeting, Winnipeg, Canada* **5**, 169.

Nielsen, C. E. (1978a). Conditions for absolute stability of salt gradient solar ponds. *Proc. ISES Congr., New Delhi* **2**, 1176.

Nielsen, C. E. (1978b). Equilibrium thickness of the stable gradient zone in solar ponds. *Proc. Am. Section ISES (Denver)* **21**, 932.

Nielsen, C. E. (1980). Chapter on non-convective salt gradient solar ponds. *In* "Solar Energy Handbook" (Dickenson and Cheremishoff, eds.). Decker, New York.

Nielsen, C. E., and Rabl, A. (1976). Salt requirements and stability of solar ponds. *Proc. Joint Conf. Am. Can. Sol. Energy Soc., Winnipeg, Canada* **5**, 183.

Nielsen, C. E., and Rabl, A. (1975). Operation of a small salt gradient solar pond. *ISES Meeting, Los Angeles, California.*

Nielsen, C. E., Rabl, A., Watson, J., and Weiler, P. (1977). Flow system for maintenance of salt concentration gradient in solar ponds. *Sol. Energ.* **19**, 763.

Nimmo, B., Dabbagb, A., and Said, S. (1981). Salt gradient solar ponds. *Sun World* 5(4), 113.

Ortabasi, U., Dyksterhuis, F., and Kaushika, N. D. (1983). Honeycomb stabilized saltless solar pond. *Sol. Energ.* **31**, 229.

Patel, S. M., and Gupta, C. L. (1981). Experimental solar pond in a hot humid climate. *Sun World,* **5**, No. 4, 115.

Payne, R. E. (1972). Albedo of the sea surface. *J. Atmos. Sci.* **29**, 959.

Pellow, A., and Southwell, R. V. (1940). On maintained convective motion in fluid heated from below. *Proc. R. Soc. (London) Ser. A.* **176**, 312–343.

Por, F. D. (1970). Solar lake on the shores of the Red Sea. *Nature* **218**, 860.

Rabl, A., and Nielsen, C. E. (1975). Solar ponds for space heating. *Sol. Energ.* **17**, 1.

Rao, S. K. (1983). "Thermal modelling of nonconvective solar pond collector/storage systems." Ph.D. Thesis, IIT, Delhi, India.

Rao, S. K., and Kaushika, N. D. (1983). Analytical model of solar pond with heat exchanger. *Energy Convers. Mgmt.* **23**(1), 23–31.

Roothmeyer, M. (1980). Soret effect and salt gradient solar ponds. *Sol. Energ.* **25**, 567.

Sargent, S. L. (1979). An overview of solar pond technology. SERI/TP-333-429, *Proc. Sol. Ind. Process Heat Conf., Oakland, California.*

Savage, S. B. (1977). A chapter on solar pond. *In* "Solar Energy Engineering" (A. A. M. Sayigh, ed.). Academic Press, New York.

Shaffer, L. H. (1975). "Viscosity Stabilized Solar Pond." U.S. Patent No. 4138,992, July 21, 1975.

Shaffer, L. H. (1978). Viscosity stabilized solar pond. *Proc. ISES Congr.,* New Delhi, India.

Shahar, S. (1968). U.S. Patent No. 337291.

Sheridan, N. R. (1981). Solar ponds in Australia. *Sol. Prog.* **9**, 2135.

Sheridan, N. R. (1982). "Solar Salt Pond at Alice Springs." Report 1382, Queensland University, Queensland, Australia.

Short, T. H., Badger, P. C., and Roller, W. L. (1978). A solar pond polystrene bead system for heating and insulating greenhouses. *Acta Horticulture* **87**, 291.

Shuman, F. (1909). *Engineering News* **61**, 509.

Sodha, M. S., Nayak, J. K., and Kaushik, S. C. (1980). Physics of shallow solar pond water heater. *Energy Res.* **4**, 323.

Sodha, M. S., Kaushika, N. D., and Rao, S. K. (1981). Thermal analysis of three zone solar pond. *Energy Res.* **5**, 321–340.

Stolzenbach, K. D., Dake, J. M. K., and Harleman, D. R. F. (1968). Prediction of temperatures in solar ponds. *Proc. Ann. Meeting SES,* Palo Alto, California.

Styris, D. L., Zaworski, R. J., and Harling, O. K. (1975). "The Nonconvecting Solar Pond: An Overview of Technological Status and Possible Pond Applications." BNWL—1891, Battelle-Northwest Laboratories, Richland, Washington.

Styris, D. L., Harling, O. K., Zaworski, R. J., and Leshuk, J. (1976). The non-convecting solar pond applied to building and process heating. *Sol. Energ.* **18**, 245.

Tabor, H. (1959). Solar collector developments. *Sol. Energ.* **3**(3), 89.

Tabor, H. (1961). "Science and New Nations." (R. Gruber, ed.), pp. 108–110. Basic Books, New York.

Tabor, H. (1963). Large area solar collectors for power production. *Sol. Energ.* **7**, 189.

Tabor, H. (1966). Solar ponds. *Sci. J.* (6), 66–71.

Tabor, H. (1975). Solar ponds as heat source for low-temperature multi-effect distillation plants. *Desalination* **17**, 289.

Tabor, H. (1980). Nonconvecting solar ponds. *Philos. Trans. R. Soc. London, Ser. A* **295**, 423.

Tabor, H. (1981). Solar Ponds. *Sol. Energ.* **27**, 181. Also published as ICAP-CATANIA SMR/87-11 as notes of Second International Symposium on Nonconventional Energy.

Tabor, H., and Bronicki, L. (1961). Small turbines for solar power package. U.N. Conf. New Sources of Energy, Rome, 35/5, 54.

Tabor, H., and Matz, R. (1965). Solar pond project. *Sol. Energ.* **9**, 177.

Tabor, H., and Weinberger, H. Z. (1980). Nonconvecting solar ponds. *In* "Solar Energy Handbook" (Kreider, J. F., and Kreith, F., eds.). McGraw-Hill, New York.

Teagen, W. P. (1973). *Proc. Solar Heating and Cooling Workshop,* Washington D.C., Part I, p. 107, March.

Tybout, R. A. (1967). A recursive alternative to Weinberger's model of solar pond. *Sol. Energ.* **11**(7), 109–111.

Usmanov, Yu. U., and Eliseev, V. N. (1973). Effect of evaporation on the thermal state of a solar pond. *Geliotekhnika* **9**, 5, 45.

Usmanov, Yu. U., Eliseev, V. N., and Umarov, G. (1971). Optical characteristics of a solar reservoir. *Geliotekhnika* **7**, 1, 28.

Usmanov, Yu. U., Eliseev, V. N., and Umarov, G. (1973). Experimental study of the removal of heat from a solar salt pond. *Geliotekhnika* **9**, 6, 23.

Veronis, G. (1968). Effect of a stabilizing gradient of solute on thermal convection. *J. Fluid Mech.* **34**, 315.

Viskanta, R., and Torr, J. S. (1978). Absorption of solar radiations in ponds. *Sol. Energ.* **21**, 17.

Wang, Y. F., and Akbarzadeh, A. (1982). A study of transient behaviour of solar ponds. *Energy* **7**, 1005.

Weinberger, H. (1964). The physics of solar ponds. *Sol. Energ.* **8**, 45.

Wilkins, E., and Pinder, K. L. (1979). Experiments with a model solar pond. *Sun World* **3**(4), 110–117.

Willsie, H. E. (1909). Experiments in the development of power from the sun's heat. *Eng. News* **61,** 511.

Wilson, A. T., and Wellman, H. W. (1962). Lake Vanda an Antarctic Lake. *Nature* **196,** 1171.

Wittenberg, L. J. (1980). "Evaluation of Solar Pond Performance." DE-AC04-76-DP00053, Mansanto Research Corporation, Annual DOE Active Solar Heating and Cooling Contractors' Review Meeting, Miamisburg, Ohio.

Wittenberg, L. J., and Etter, D. E. (1982). Heat extraction from a large solar pond. ASME-82-WA/Sol-31.

Wittenberg, L. J., and Harris, M. J. (1980). Management of a large operational solar pond. *Proc. 15th Intersoc. Energy Conv. Eng. Conf.,* Seattle, Washington, p. 1435.

Wittenberg, L. J., and Harris, M. J. (1981). Construction and start up performance of the Miamisburg salt-gradient solar pond. *J. Sol. Energy Eng.* **103,** 11.

Zangrando, F. (1979). "Observations and Analysis of a Full Scale Experimental Salt Gradient Solar Pond." Ph.D. Thesis, University of New Mexico, Albuquerque, New Mexico.

Zangrando, F. (1980). A simple method to establish salt gradient solar ponds. *Sol. Energ.* **25,** 467.

Zangrando, F., and Bryant, H. C. (1977). Heat extraction from a salt gradient solar pond. *Proc. Int. Conf. Alternative Energy Sourc., Miami Beach,* Florida.

Zangrando, F., and Bryant, H. C. (1978). Solar ponds. *Sol. Age* **3,** 21.

The Demand for Home Insulation: A Study in the Household Demand for Conservation

J. Daniel Khazzoom

Department of Economics
University of California
Berkeley, California

I. BACKGROUND

A. Motivation of the Study

There has been a growing need over the past few years for the development of a behavioral model of the demand for conservation. This happened as the electric utilities, as a group, became increasingly a supplier of conservation in the same sense that, as a group, they are a supplier of energy. Before this change, utilities needed to estimate the future demand for energy only in order to ensure a balance between the supply of energy and the expected demand for energy in the future. But the new responsibilities confronting them meant that, just as in the case of energy demand, they needed to know how the demand for conservation changes as the economic conditions change if they are to ensure a balance between the supply of conservation and the demand for conservation in the future. Hence the need for a behavioral system that can capture the relationships of interest.

But the interest in behavioral modeling of the demand for conservation was also heightened by the realization that a utility can enhance the allocation of its investment resources by opting for an active role in shaping the demand it faces—for both energy and conservation— rather than playing a passive role of merely responding to changes in outside conditions. (The activist view of the firm is not unknown in the economic literature, although it has been swamped by orthodox theory. Economists like Clark (1955), Galbraith (1967), and Schumpheter (1934) have stressed the activist role of the firm. Several prominent critics of orthodox theory focused their criticism on the passive nature of the firm depicted by the theory. They argued that in most industries firms try to modify the demand for their product rather than merely react to market conditions.) The new emphasis may have been prompted by the investment opportunities that conservation opened up. It could also have been strengthened by the awareness that since conservation demand and energy demand substitute for and interact with each other extensively, no investment program can be optimal if it explores the investment opportunities in one (energy) but ignores the opportunities in the other (conservation).

The development of a model which estimates behaviorally the demand for conservation (and which can also be easily integrated with a model of energy demand) provides utilities and the regulatory agency with a tool that makes it possible to estimate the response of the household to measures intended to affect conservation demand one way or another. These measures may originate with the utility (e.g., subsidized interest payment, rebate, higher energy prices) or with the state and federal government (e.g., tax credits).

While our discussion so far has centered around the electric utilities, the interest in modeling conservation demand is not confined to electric utilities, but extends to major sectors of the energy industry—gas utilities, pipelines, oil companies, coal companies, etc.—as well as to energy researchers in universities and policy makers in government. This study is the first step toward providing us with the capability of estimating one major ingredient of the demand for conservation, namely, the demand for insulation. To establish feasibility, the study focused on a relatively small service area, the Sacramento Municipal Utility District (or SMUD) service area. But the model can be extended without difficulty to larger geographic areas.

The study focuses on home insulation rather than home thermal integrity because during the sample period changes in home insulation were practically the only source of change in the thermal integrity of homes in the Sacramento area. But the methods developed in this study to generate the indices of insulation demand, as well as the econometric methods used in estimating this model, can all be used directly to estimate the demand for any combination of measures (other than just home insulation) that affect the thermal integrity of the home. Examples: double glazing, installation of hangovers, changing the direction of windows away from, or toward, the sun, and so on.

B. Plan of the Work

Section I provides the background material for the study. It also introduces the reader to the characteristics of the service area for which this model is estimated and explains briefly the strategy we followed in this study for choosing the explanatory variables. Section II discusses the specification of our model, drawing on theoretical as well as institutional information. Section III introduces the novel concept of the index of insulation for homes of various types. The index is based on the output of an engineering model of house characteristics. It (the index) provides the link for integrating the engineering and economic approach to modeling the household demand for conservation and energy. Section III also explains how the index is constructed and how an unbiased aggregate index can be derived for any desired level of aggregation.

Section IV discusses problems that logically precede the estimation step. Our concern here is motivated partly by the question of how one can estimate a model without having to impose prior restrictions on the form of the relationship used in estimation. Here we discuss the Box–Cox transformation and summarize recent results on this transformation. We also discuss the controversy of marginal (versus conditional) variance estimation with the Box–Cox transform and address problems that the applied economist encounters in the use and interpretation of marginal estimates. Section IV also addresses the problem of simultaneity that arises in our model.

Section V reports estimation results using an iterative ordinary least-squares procedure (IOLS) and explores the historical behavior of some elasticity measures. We also report results based on estimates of the marginal variances and discuss scaling and numerical problems we encountered. We also report estimation results using bootstrap methods to assess the variability of our estimates.

In Section VI we turn to consistent estimation. The use of consistent methods of estimation is necessary because insulation demand is jointly determined with energy demand. We first extend the limited information maximum likelihood method (LIML) to a Box–Cox transformed relationship which also contains more than one endogenous variable. We report LIML results for our model and then subject our model to several tests. We also estimate by using a nonlinear two-stage (NL2S) least-squares procedure. (We found that in general this estimator does not exist for a Box–Cox transformed model.) We conclude by adopting model 5* of Table XXIV to approximate the household demand for home insulation.

Section VII is a summary.

C. The Utility

The Sacramento Municipal Utility District is a publicly owned municipal utility. It was created on July 2, 1923, under the California Municipal Utilities District Act by a vote of the electorate. It is governed by a board of five directors elected by ward for staggered four-year terms. The Sacramento Municipal Utility District sells electric power only, and it is the sole distributor within an area of about 750 square miles in central California. Its service area consists of a major portion of Sacramento County and a small adjoining portion of Placer County. The service area has a population of over 760,000.

Table I shows the average number of residential customers in 1960–1980 in SMUD's service area as well as the estimated distribution of these customers in 1960–1980 by dwelling type: single-family, duplex, multiple-family, and mobile homes. Table II shows the breakdown of customers by energy source used for space heating. The growth of the number of customers with electric heat is remarkable. Between 1968 and 1981 the compounded growth rate was in excess of 22%. In no year was the growth rate smaller than 11%. On the other hand, gas-heated homes grew at an average rate of 2.3% annually during the same period. Hence, households with electrically heated homes managed to increase from 2.4% of the total number of customers in 1968 to 20.3% of the total number of customers in 1981. Table III shows the break-

TABLE I

Average Number of Residential Customers in SMUD's Service Area and Estimated Distribution of SMUD's Customers by Dwelling Type[a]

Beginning of year	Number of residential customers	Estimated distribution of customers			
		Single family	Duplex	Multiple family	Mobile
1960	138,476	109,897	6,172	19,586	2,821
1961	146,033	115,930	6,507	20,610	2,986
1962	151,560	119,977	6,792	21,684	3,107
1963	158,293	123,903	7,403	23,733	3,254
1964	164,993	126,027	7,897	27,664	3,405
1965	170,240	128,979	8,518	29,222	3,521
1966	175,433	132,552	8,871	30,378	3,632
1967	179,959	135,928	9,136	31,164	3,731
1968	184,051	138,498	9,311	32,308	3,934
1969	188,666	141,057	9,581	33,889	4,139
1970	195,249	144,256	9,890	36,770	4,333
1971	203,312	146,606	10,112	41,682	4,912
1972	213,515	149,722	10,692	47,350	5,751
1973	222,607	150,690	11,285	53,770	6,862
1974	230,920	152,985	11,759	58,376	7,800
1975	239,293	157,279	12,420	61,146	8,448
1976	247,644	160,663	13,271	64,805	8,905
1977	258,572	166,305	15,266	67,315	9,686
1978	270,500	172,130	17,620	70,403	10,347
1979	283,598	179,899	18,924	73,764	11,011
1980	294,946	186,765	19,901	76,666	11,614

[a] Source: Total number of residential customers is derived from SMUD's files. Breakdown by type of dwelling: 1960: County data used in calculating the numbers of the four dwelling types in Sacramento County are derived from Bureau of the Census, *U.S. Census of Housing, Final Report HC (1) 1*, Table 17, pp. 1–111. The data are converted to SMUD's service area data by multiplying by 0.94032. 1970: County data used in calculating the number of each dwelling type in Sacramento County are derived from Bureau of the Census, *Census of Housing: Housing Characteristics for States, Cities, and Counties Vol. 1, Part 6, California*, Table 48, pp. 6–310. County data are converted to SMUD's service area data by multiplying by 0.94032. 1980: County data used in calculating the number of single-family and mobile homes in Sacramento County are derived from California DOF internal files. County data used in calculating the stock of duplexes are derived from estimates made by California's Department of Housing and Community Development. County data used in calculating estimates of multiple-family units are derived from the difference between DOF's internal estimates of structures with two or more units and DOH estimates of duplex units. County data are converted to SMUD's data by multiplying by 0.94032.

TABLE II

Number of Customers in SMUD's Service Area[a,b]

| Beginning of year | Customers with electric heat | | Customers with gas heat | | % of customers with electric heat: $(1) \div [(1) + (3)]$ |
	Winter (1)	Summer (2)	Winter (3)	Summer (4)	(5)
1968	4,389	4,437	179,879	179,181	2.4
1969	5,562	5,554	183,343	182,632	2.9
1970	7,308	7,581	187,899	187,750	3.7
1971	10,088	10,199	193,265	193,033	5.0
1972	14,084	14,760	199,144	199,307	6.7
1973	18,540	18,754	204,070	203,849	8.3
1974	22,353	22,517	208,564	208,409	9.7
1975	25,744	25,892	213,488	213,522	10.8
1976	28,906	29,253	218,689	218,490	11.7
1977	34,167	34,632	224,207	224,337	13.2
1978	40,920	41,254	229,808	230,264	15.1
1979	48,380	49,201	234,954	234,922	17.1
1980	55,378	55,965	241,658	239,276	18.8
1981	61,723	62,293	241,477	241,426	20.3

[a] Source: SMUD's Files.
[b] Summer covers June–September. Winter covers the balance of the year.

down of electricity sales to electrically heated and gas-heated homes during the summer and winter in 1968–1981. Much of the growth occurs in the sale of electrically heated dwellings.

An overview of SMUD's insulation program is reported in Khazzoom (1984). For the details of the utility's insulation programs, see Sacramento Municipal Utility District (1980, 1981, 1982).

D. Strategy for Selecting the Design Matrix

Bearing in mind that the motivation behind model construction is the need to provide an approximation of reality, the basic model requirement is simplicity; that is, it should contain a small number of parameters. This is often referred to as the principle of parsimony. At the same time we want the model to be able to capture the essence of the phenomenon we are investigating, to extract, so to speak, the main features of the process that is being studied. This entails the identification

TABLE III

SMUD's Electricity Sales in kW h^{a,b}

	Electrically heated homes		Gas-heated homes	
Year	Winter	Summer	Winter	Summer
1968	69,750,765	18,021,127	716,779,341	396,814,655
1969	85,109,717	22,660,470	794,705,598	450,646,937
1970	95,070,211	29,532,008	815,187,084	496,513,244
1971	135,383,308	39,110,082	904,836,398	530,494,448
1972	165,742,359	50,740,580	938,230,903	575,798,788
1973	202,729,440	63,514,544	977,244,712	627,399,249
1974	220,843,874	74,637,103	984,204,559	621,499,517
1975	285,432,074	83,182,658	1,041,745,949	620,941,531
1976	286,701,071	96,079,684	1,071,894,369	645,164,021
1977	329,008,184	118,483,621	1,091,296,419	680,895,932
1978	408,994,537	150,643,058	1,155,968,240	739,310,981
1979	523,254,626	185,252,172	1,213,313,527	752,651,533
1980	553,147,808	193,812,723	1,179,445,739	656,389,246
1981	586,737,670	243,453,794	1,192,268,441	767,604,951

^{a} Source: SMUD's files.
^{b} Summer covers June–September.

of the key variables that propel the phenomenon under investigation. Finally, we want the model to pass statistical tests reasonably well.

The problem of model selection strategy is prompted by these requirements and stems primarily from the fact that we do not know with certainty the "appropriate" choice of the column dimension of the design matrix (the matrix of explanatory variables). Concomitant aspects of the same problem are our uncertainty about the timing of these variables, their interrelationships, as well as the functional form under which they enter the relationship. (We discuss the choice of functional form separately later in Section IV.A.)

The problem may be posed in the following general terms: We want to estimate the parameters of a model that are known to lie in a K-dimensional parameter space. However, we suspect that these parameters lie close to a lower-dimensional parameter space K_1 in the sense that the extra variables $K_2 = K - K_1$ do not contribute "substantially" to our identification of the ingredients of the economic processes by which our data are generated. Seen in this context, the problem boils down to a trade-off between bias and sampling variability. We might very well be able to estimate the parameters in the higher K-dimensional space. They will be unbiased. Unfortunately, they will generally have large variances. On the other hand, if the estimates are restricted to the lower K_1-dimensional space, they will generally have smaller variances. But their biases may be large, all depending on the quality of our prior information.

In the context of our model, the general design matrix may be partitioned, at least conceptually, into two disjoint sets: (1) Z_1, which contains explanatory variables that are important in determining the demand for insulation, and (2) Z_2, which contains explanatory variables that have some impact on the demand for insulation. Our problem is that we do not know with certainty which variables fall in each set. The task of selecting the "appropriate" set of explanatory variables breaks down to finding out of the total set $Z_1 U Z_2$: (1) which variables make up the elements of Z_1; (2) which elements of Z_2 should we include in the model and which elements we should exclude. Of these two tasks, the second is the hardest.

Uncertainty about the appropriate dimension of the parameter space means that the investigator has a choice of one of two strategies.

1. Start with a very general specification and test hypotheses in increasing order of restrictiveness until a more specific model is obtained. The main feature of this approach is the abandonment of any attempt to identify a suitably parsimonious model at the outset. Given a set of possible model specifications, a structure is imposed on the problem by nesting the hypotheses to be tested, i.e., by making each specification a subset of another more general specification. These hypotheses form a uniquely ordered set which enables the researcher to carry out a sequential testing procedure down to the more restrictive specification. Once a null hypothesis is rejected, all succeeding hypotheses are rejected as well, and the testing stops. When the error terms are normally distributed, it is common practice to employ the classical testing procedures, particularly the likelihood ratio (LR) test. However, LR requires that the model be estimated under both the null and alternative hypotheses. In the event that the estimation of the restricted model is computationally too difficult or costly to carry

out, a Wald (W) test may be used. A Wald test requires the estimation of the unrestricted (the more general) model only. Alternatively, if the estimation of the unrestricted model is costly, one may use the Lagrange multiplier (LM) test, which requires the estimation of the restricted model only. (For the case when estimation is difficult even under the null hypothesis, Breusch and Pagan (1980) proposed a pseudo-LM statistic based on the $C(\alpha)$ statistic proposed initially by Neyman. It involves the construction of a pseudo-LM test with asymptotic properties identical to the true one.)

2. Start with the most restrictive model and test for misspecification. Unlike a specification test, a test of misspecification has, in principle, no clear alternative, and so it usually involves one procedure or another for assessing the goodness of fit by the maintained hypothesis. However, this is not always the case. A misspecification test may be motivated by a suspected departure from the maintained hypothesis in some particular direction. In the case in which we start with the restrictive model, as in the case of our model, the alternative is that some variables that have been left out should have been included. Under the assumption of normality, any of the classical tests LR, W, and LM can be used.[1]

Other than the tests just cited, the literature contains several criteria and empirical rules for choosing between bias and variance. Some, like the classical tests, and Theil's criterion of maximizing $(\bar{R}^2)$ are not based on a decision theoretic context in the sense that they ignore the loss function. Others such as Mallows's Cp, Akaike's information criterion (AIC), and Amemiya's prediction criterion (PC) include a consideration of the loss function associated with choosing an incorrect model. [For a review of these and other single-statistic criteria as well as a discussion of the relationships among them, see Amemiya (1980)

---

[1] When the log likelihood is quadratic, the three tests give identical results. See the excellent exposition in Buse (1982). Otherwise, they are only asymptotically equivalent, and as has been often pointed out in the literature, a systematic numerical inequality exists between the three when used to test linear restrictions on the coefficients of certain linear models. The result is that these large sample tests can yield conflicting results. Evans and Savin (1982) investigated the power of the large sample tests and found that the nominal significance level was very poorly approximated—which explains why the power functions of the large sample test differ so markedly from the power functions of the exact tests. The authors adjusted the critical value of the asymptotic distributions and found that the probability of conflict dropped substantially as a result.

and Breiman and Freedman (1983). See also Judge *et al.* (1980).] All of these criteria involve error comparison. However, not all criteria presume normality of the error term. But when normality holds, all of these criteria are translatable into F tests.

Of the two strategies for selecting the regressors, the forward selection of variables has obvious computational and cost advantages. However, the test results in general will not be independent. In contrast, the strategy of backward elimination of variables has, at least in principle, optimal properties which follow from the independence of the tests. However, the greater the number of times the tests are conducted before a final model is chosen, the greater the size of the test becomes.[2]

Properly done, the backward elimination requires the specification of the set of all possible regressors. This is usually an impossible task, since one is never sure of such a set. Willy-nilly, the backward elimination strategy will usually involve some forward selection of variables as well. At any rate, it is doubtful that the optimality properties of the backward elimination strategy, even if applied strictly, do really hold. Both strategies make the model (and the estimation procedure) dependent on the results of hypothesis tests based on the data at hand. The sampling properties of these estimators (often referred to as pretest estimators) are usually unknown. This limits our knowledge of their (the pretest estimates') inferential reach. And in a decision-theoretic context, we are really uncertain of the risk associated with the action taken on the basis of these estimates.

In short, the problem of the correct way of going about choosing the column dimension of the design matrix is unresolved. This is not a satisfactory state of affairs but is not unexpected, given the complexity of the problem. Under the present state of the art, the solution will have to rest on pragmatic grounds.[3]

In this study, our approach is based on identifying, at the outset, a suitably parsimonious model by drawing as much as possible on all the

---

[2] Some economists (see, e.g., Sargan, 1980) favor the backward elimination strategy when the model consists of a stochastic difference equation, since this approach helps identify common factors in the lag structure. The problem did not turn out to be an issue in our model.

[3] Summarizing the problem of criterion choice, Amemiya notes "that all of the criteria considered are based on a somewhat arbitrary assumption which cannot be fully justified, and that by slightly varying the loss function and the decision strategy, one can indefinitely go on inventing new criteria.... Thus, ... the selection of regressors should be primarily based on one's knowledge of the underlying economic theory and on one's intuition which cannot be easily quantified." See Amemiya, 1980, p. 352.

available *a priori* information—theoretical as well as institutional—to narrow down the range of admissible specification of the design matrix. To do that, we went through two steps. First, we identified those variables that theory suggests are members of $M = (Z_1 U X_2)$, where $X_2 \subset Z_2$ includes variables that theory suggests may play some relatively important role in determining the demand for insulation. Here M is a relatively large set in the sense that it contains a large number of elements (12 variables in our case). Its size reflects the extent of our uncertainty, since no theoretical work has been done on the household demand for insulation. Second, to eliminate those variables in M whose *quantitative* effect is secondary, we turned for clues to survey results available for some California utilities and other utilities outside California. (Several utilities have commissioned follow-up surveys to investigate the household's motivation toward conservation measures, but we focus primarily on the follow-up results of Pacific Gas and Electric's (PG&E's) ZIP program, which as evidence suggests was used by customers primarily to finance insulation. See Pacific Gas and Electric Company, 1982b, p. II-5.) To the extent that the respondents identified an element in M as being of secondary importance in their conservation decision, we dropped that variable from M.

Since factors affecting conservation decision may differ according to the measure in question, it is conceivable that a variable that was not significant in the demand for some conservation measures may, in fact, be important in the demand for insulation. Whenever we suspected that this may have occurred, we sought further confirmation from PG&E's ZIP follow-up surveys before we made a decision to drop a variable from M. Where such a confirmation was not possible, we did not drop the variable from M. (An example is the effect of the oil crisis on the demand for insulation.) This left us with a smaller set $X \subset M$. Since M, as it turned out, included all variables identified by one survey or another as the one or two most important variables in the demand for conservation, the set does not leave out variables that appear in the union of key variables in these surveys.[4] When there was a lingering doubt about variables that were not included but that could

---

[4] A test of the model's estimates based on the restricted set X of explanatory variables does not yield self-fulfilling results, since our sample period and the period of the follow-up surveys coincide at most in two years. Our sample period covers 1969–1980, while the follow-up surveys relate mostly to 1980 and at most 1979–1980. As such our results yield implicitly a backcast test of the hypothesis that the key factors that affected insulation demand in 1980 or 1979–1980 were also the major factors that affected insulation demand in the preceding 10 years.

possibly belong (we could think of only one such variable, the cost of insulation), we tested subsequently for misspecification.

II. SPECIFICATION

A. Determinants of the Demand for Insulation

The demand for insulation originates primarily from existing homes (retrofitting) and partly from new construction. Factors that affect the demand in one market are likely to affect the demand in the other. In this study we will be looking at the aggregate demand originating in the two markets combined. The analysis may be further refined and the demand can be estimated in each market separately.

In general we expect factors that increase the attractiveness of insulation to increase the demand for insulation; and vice versa, factors that reduce the household interest in insulation (e.g., by increasing the opportunity cost of insulation) may be expected to decrease the demand for insulation. In a similar manner, we expect factors that increase the awareness of the benefit of insulation to lead to a step-up in the demand for insulation; and vice versa, factors that increase the awareness of the adverse effects of conservation may be expected to lead to a reduction in insulation demand, other things being equal.

We turn now to a more specific discussion of these factors.

1. Factors Increasing the Demand for Insulation

a. Energy Price. In as much as a higher energy price enhances the value of the electricity (or gas) saved, to that extent an increase in energy prices may be expected to increase the demand for insulation and vice versa when energy prices decline. Since in SMUD's service area only electricity and gas are in use (and will be in use for the foreseeable future), the two relevant energy prices are electricity and gas prices. But they are not equally important for residential customers. For an all-electric or an almost-all-electric home, the relevant price is electricity's.[5] For gas-heated homes, both prices should, in

[5] The split of homes between gas-heated and electricity heated homes must logically take into account both the price of electricity and the price of gas. But once a customer has opted for electric heat, gas price becomes practically irrelevant, except for gas water heaters and gas driers. In general, the energy used by the gas water heater (and gas drier) is not affected by the insulation level. Only if the household changes the thermostat setting after insulation will the amount of gas used for water heating be affected.

principle, enter the relationship. However, to the extent that space heating is the propelling force behind insulation demand, we may expect gas price to be the prime mover in determining the demand for insulation by gas-heated homes.

b. Size of Energy Usage by the Household. Logically, we expect the kilowatt hours used for space heating and air conditioning to be an important determinant of the demand for insulation in electrically heated homes. In general, the greater the usage is, the greater is the benefit that accrues from insulating, and hence the greater is the demand for insulation.[6] From a practical standpoint, since space heating and air conditioning are major constituents of the winter and summer demand for electricity in electrically heated dwellings,[7] we may expect a similar, although perhaps a weaker relationship to exist between the total electricity use and the demand for insulation.

The relationship between electricity use and the demand for insulation is likely to be weaker for gas- than for electrically heated homes. This is so because the resulting saving in electricity will be larger in an electrically heated home than in a gas-heated home. Indeed, for gas-heated homes, both gas and electricity usage are, in principle, relevant: gas usage in the winter and electricity usage in the summer. But if we are to choose only one of the two energy sources used by gas-heated homes, gas usage is likely to be the one that provides the impetus for insulation demand. However, our experience in this field is limited and the available information is fragmentary. But the data available (including engineering results used to calculate the insulation indices) suggest that the British thermal units used for space heating are much larger than the British thermal units used for air conditioning.

---

[6] Strictly the statement is correct only if we hold the level of insulation and area constant because of the decreasing marginal productivity of insulation. The first conditioning (on insulation level) poses no problem for the aggregates. As to the area, for electrically heated homes, the area changed very little between 1969 and 1980. (See column 10, Table VIII below.)

[7] Time series of these shares are not available for SMUD's service area. Metering records of electricity used for space heating in the early 1970s in a sample of electrically heated SMUD customers were, unfortunately, destroyed. For 1969, however, estimates show that electric space heating accounted for 35% of the total electricity consumed during the eight months of January–May and October–December 1969. For gas-heated homes, the comparable air conditioning percentage was 36% in 1969. See Table 6-2 in Khazzoom (1986).

Two points before we go further:

1. Unless the share of electricity (or gas) used for space heating remains a constant in the total electricity (or gas) usage in the winter, estimating the insulation in demand with the total (rather than the space heating) usage will be in the nature of a second best. Similar remarks apply to total electricity usage as opposed to air conditioning usage in the summer. But generally homes are not metered separately for air conditioning and space heating, and, as footnote 7 makes clear, whatever sampling results were available for space heating usage have been destroyed. Empirically we have no choice with the data to be used in estimation. From the econometric point of view, this raises a simultaneous-equation problem. We discuss consistent estimation in Section VI.

2. When we talk about electricity (or gas) saving, we are confining our remarks to the benefit that results from the mechanical effect of insulation. That is, other things being equal, the installation of a higher level of insulation reduces the demand for energy. We do not imply that other things will indeed remain equal or that the benefit of insulation will necessarily be maintained over time. Indeed much of the benefit of insulation is likely to be eroded over time (through the effect on the price level). (See Khazzoom, 1986, Table 10.4.)

2. Factors Decreasing the Demand for Insulation

a. Interest Rates. Inasmuch as the demand for insulation is a demand for a durable, we would expect higher levels of interest rate to have an adverse effect on the demand for insulation.

b. Cost of Insulation. Here again as in the case of the demand for a durable, we would logically expect the cost of insulation to have a negative impact on the demand for insulation. But the analogy with a durable should not be carried too far. Insulation, unlike a tangible durable, is not a fixed object but a mixture of attic, wall, and floor insulation, with a varying mixture of the three. It is well known, for example, that the marginal productivity of increased attic insulation approaches zero asymptotically when the resistance level increases beyond R-11. In that range, the marginal cost of a kilowatt hour saved increases rapidly. But the consumer is not locked in to insulating the ceiling only. As long as the resistance level of wall insulation is at, say, R-7, the marginal productivity of insulating the wall will be larger than the marginal productivity of insulating the attic. Hence the household can lower the marginal cost of a kilowatt hour saved by insulating the wall rather than the ceiling. Similar remarks apply to floor insulation. In general an informed household will aim at that insulation level at which the marginal cost of ceiling, wall, and floor insulation is equal. Because insulation is likely to proceed in discrete steps, there will be at any time large differentials in the marginal costs and therefore a large potential for marginal cost reduction by switching from attic to wall insulation. The possibility exists that the consumer may not view the cost of insulation as a deterrent because of his ability to skirt cost increases by switching from ceiling to wall or floor insulation. Empirically this means that the relationship between the cost of insulation and the demand for insulation may be erratic.

At any rate, since the historical experience of home insulation does reflect this switch between these various types of insulation, the operational definition of cost should reflect the same type of switch back and forth between ceiling, wall, and floor insulation.

Conceivably, the interest rate, the insulation cost, and the depreciation rate of installed insulation enter the relationship as one combined variable in the form of a user's cost. But we doubt that. First, the depreciation rate is about the same for ceiling, wall, and floor insulation, so that the depreciation rate simply adds a constant to the interest rate. Second, it is very doubtful that households do indeed think in terms of user's cost. And the efforts to incorporate a user's cost variable in energy-related models (such as appliance demand) have generally produced mixed results. (See, for example, Taylor *et al.,* 1982, Table 7-5, p. 7-9.) Nonetheless, we will test for the effect of including a cost-of-insulation variable in the model.

c. Income. In general we may reasonably expect insulation to be an inferior good. This is likely to be the result of the combined effect of two factors: (1) the decline in the marginal utility of income as income rises, at least beyond a certain level (this is another way of saying that the marginal utility of the saving that results from increased insulation declines as income increases)[8] and (2) the tedious nature of the insula-

---

[8] The marginal utility of income declines when the utility function is concave. Concavity is a stronger assumption than the quasi-concavity assumption one normally makes in

tion process. Insulation does not come as easily as flipping on the switch or turning on the furnace. It imposes requirements that generally fall outside the range of the daily routine of most households. One has to compile a list of contractors, which most households are not adept at doing; stay at home during the contractors' visits; make sense of the differences in the work specifications given by each contractor; screen the reputations of the bidding contractors; and finally stay at home during the period when the insulation work is done. The process may take days, and it may be quite exacting. For this reason, and coupled with the declining marginal utility of income, it is not unreasonable to expect an increase in the level of household income to weaken the household's motivation to insulate and to exert a downward pressure on the demand for insulation. (See also the discussion in Section V.A.)

d. Intangible Costs. These are cost items that are not quantifiable in an objective way but that nonetheless affect the household's demand for insulation adversely. Their impact depends on the subjective evaluation of the household. Examples are the time required to do the insulation on a do-it-yourself basis, the risk of letting in an (unknown) contractor to do the work in the house, etc. Although the valuation of intangible costs will vary from one household to another, it is probably not unreasonable to expect it to be directly related to the income level (that is, the opportunity cost of the time spent interacting with contractors increases with the increase in the income level) and therefore to be subsumed by the income effect just discussed.

3. Factors Increasing the Awareness of Insulation

a. The Oil Crisis. The large increase in oil prices in late 1973 and continuing in 1974 has perhaps done more than anything else to heighten the public's awareness of the advantages of conservation in general and perhaps to jolt most households into a stronger response to higher energy prices than had been the case until then.

b. Utility Initiative. The utility initiative is reflected in the efforts made by SMUD, PG&E, and various other agencies to provide infor-

---

utility theory and is perhaps too restrictive for theoretical purposes. But it is generally used in applied work and is probably not unreasonable in the context of the income levels in Sacramento County.

mation and to promote conservation programs inclusive of insulation. At SMUD the first attic insulation program was initiated in the second half of 1977. At PG&E (which services the gas-heated homes in SMUD's service area), the major effort to promote insulation measures began at about the same time.

c. Ownership of Efficient Appliances. Ownership of efficient appliances may be taken generally as indicative of an awareness of the advantages of conservation. But aside from that, its impact on the demand for insulation is uncertain. On the one hand, it may be a factor that increases the relative attractiveness of insulation and hence leads to an increase in insulation demand. On the other hand, the ownership of more efficient appliances may proceed simultaneously with increased insulation. And so the contemporaneous ownership of more efficient appliances may not necessarily imply the existence of a greater saving potential left in untapped insulation.

d. Experience with the Harmful Side Effects of Insulation. Experience with the harmful side effects of insulation is likely to decrease the demand for insulation. Much has been said, for example, about the effect of hermetically insulating homes on indoor pollution. In that context, it has also been pointed out that the household's effort to increase ventilation (to combat the increased indoor pollution) has usually resulted in the loss of a good deal of the energy saved by insulation. But no conclusive quantitative study has been done on this subject.

4. Other Factors Affecting the Demand for Insulation

a. Portability. Other things being equal, the more portable is the conservation measure, the greater one would expect its demand to be. Insulation is not portable at all. Consequently the duration of one's stay in a certain dwelling will influence one's interest in insulating the home. In general, we expect an increase in the proportion of tenants in the total population to exert a downward pressure on the demand for insulation. We also expect a negative relationship to exist between age and the demand for insulation for similar reasons.

b. Type of Home. Single-family homes generally have an attic or a crawl space, while multiple-family and mobile homes typically may not. To the extent that such differences in space design are widespread, we expect an increase in the share of single-family dwellings in the total dwellings to be associated with an increase in the demand for insulation and vice versa. Moreover, single-family homes tend to be larger than multiple-family homes and mobile homes. We therefore expect the use of electricity in single-family homes to be, in general, larger than in other homes. By our earlier analysis, this exerts an upward pressure on the demand for insulation.

In short, on both counts, an increase in the share of single-family homes will create conditions that are conducive to a greater demand for insulation. Operationally it will not be possible to separate out the effect of the type of dwelling from the effect that operates through the level of energy use. Moreover, we are not certain how significant is the difference in the construction design between single-family homes and other dwelling types as far as the existence of a crawl space in the attic is concerned.

5. The Adjustment Process

In discussing the role of energy consumption in the demand for insulation, we concluded that, other things being equal, the higher the energy consumed by the household, the greater the expected demand for insulation. Other things are not equal, however. To the extent that the household has already insulated to the desired level, we do not expect the existing level of consumption, high as it may be, to lead to any further increase in the demand for insulation. Only when the adjustment to the desired level has not been completed do we expect a further movement toward increased insulation level. Similar remarks apply to the price variable. Higher prices will tend to induce a greater demand for insulation. However, to the extent that the target level of insulation is not achieved within the year, further movement toward increased insulation will take place in later years, even if the price level remains constant.

From all we know, it is reasonable to assume that the impact of a change in one of the determinants of the demand for insulation will only partially materialize within the year during which the change occurs. By the same token, it is equally reasonable to conclude that as long as a discrepancy exists between the target level and the existing level of insulation, the historical change in the determinant will continue to affect demand, even if the determinant remains constant after the initial change.

These remarks suggest that the dynamics of the household demand for insulation may be well approximated by a partial adjustment model. This approximation is also consistent with the view that the historical experience rather than just the present levels of prices, energy usage, income, etc., play a role in determining the demand for insulation.

B. Lessons from the Utilities' Follow-up Surveys

Several California utilities commissioned follow-up surveys for the Residential Conservation Service (RCS) audit and utility conservation programs. We use follow-up surveys to help us narrow the specification of our model. Basically, we search for clues in these surveys to help us identify (from among the theoretical variables suggested by our analysis in the preceding section) those variables that are quantitatively significant in determining the demand for insulation.

Observe that we do not expect a one-to-one correspondence between audit (or conservation measures in general) on the one hand and insulation on the other—although a rough estimate from Pacific Gas and Electric (PG&E) indicates that on the average a hundred audits result in the installation of attic insulation for about fifteen of the households that asked for an audit. But there are other problems in the survey results. One is the discrepancy between the ex post and the ex ante demand for audit. In California as a whole, 5% of the sampled customers who were listed among those who did not call for an audit actually did call for an audit, but the utility did not, or could not, respond to their call for one reason or another. Moreover, there is a great deal of variability in this percentage. In PG&E's service area, such customers accounted for 10% of the sampled customers listed among those who did not call for an audit (Pacific Gas and Electric Company, 1982a, p. IV-31). Similarly the decision to insulate or not to insulate cannot be attributed solely to the utility's efforts. There are numerous other media that enhance the household's awareness of the benefits of conservation and provide information on various conservation programs including insulation.

But we also note that while it is true that utility programs are not the only media, none of these media, including the utility's audit or PG&E's ZIP program, are the propelling force behind insulation. They provide necessary but not sufficient conditions for insulation. In short, they do not operate in a vacuum. Rather, they build on demand created by economic circumstances, such as an increase in the cost of energy or a reduction in income. But by telling us why customers did or did not choose to call for an audit or to follow up on an audit recommendation, these surveys help unveil the key factors that influence the household's demand for insulation.

We reiterate at this point what we said in Section I.D: In going through the survey results and in translating their implications for insulation demand, we pay special attention to PG&E's ZIP follow-up program, since evidence shows that this program was used by households mostly to finance insulation (Pacific Gas and Electric Company, 1982b, p. II-5).

We turn to the major findings of these surveys.

1. Survey Findings

a. Energy Prices. Many respondents did not specifically list the price of energy as a key factor that affected their interest in audit and conservation. But this factor was implied in various ways, almost taken for granted, in the participants' answers. Many did list rising energy prices specifically as the reason behind their request for an audit or for their demand for specific conservation measures. (See also the discussion in Weiss and Newcomb, 1982, p. 10.)

b. Energy Use per Household. All surveys found that the participants in conservation programs, including the ZIP program, tend to have a much higher energy use per household[9] than nonparticipants. (See, for example, Pacific Gas and Electric Company, 1982a, p. V-18.) Not surprisingly, higher energy use per household has been found to be associated with demographic and social characteristics which occurred in many survey results: higher income for the household, larger than average family size, greater square footage of the residence, and greater likelihood that the dwelling is a single-family type rather than a multifamily or duplex type. These characteristics logically tend to go hand in hand with a larger average energy usage for the household.[10]

---

[9] There is evidence that between electrically and gas-heated homes, the heavy users of electricity are more likely to participate than the heavy users of gas. This may be because gas prices had not yet escalated at the time of the survey to a level commensurate with the rate of increase in the price of electricity.

[10] As noted in the text, these characteristics appeared in many surveys. For example, Edison's SAVE II follow-up survey found that the household size of the participants was not significantly different from the household size of the nonparticipants (Southern California Edison, 1982, 4, XIII-4 to 6).

188 J. Daniel Khazzoom

Two observations are in order at this point.

1. It is important to distinguish between the *indirect* role of higher income level as a characteristic associated with greater energy use and the role of higher income as a factor that affects *directly* the demand for insulation. Higher income level increases the consumption of energy. Hence we would expect higher income to be observed whenever we observe higher energy consumption per household, other things being equal. Since higher energy use increases the demand for insulation, the *indirect* effect of an increase in income on the demand for insulation is positive. On the other hand, the *direct* effect of increased income may be reasonably expected to be a downward pressure on the demand for insulation, as we discussed earlier. The net impact of the direct and indirect effects may be positive or negative. The net impact can be calculated from the solved reduced form of a jointly estimated model of the demand for electricity, the demand for insulation, and the demand for efficient appliances. As it is, the survey results are generally not explicit on the direct effect of income.

2. A similar remark applies to the *type of home*. We observed earlier that energy use in single-family homes is generally greater than the use in multiple-family or duplex units because single-family homes tend to be larger. On the other hand, the single-family dwelling may also have an opposite effect, because single-family homes have attic or crawl spaces (which provide the prerequisite for ceiling insulation), which multiple-family and duplex units do not. (See also the discussion in Pacific Gas and Electric Company, 1982a, p. VII-10.) But the evidence from these surveys, although indirect, casts doubt on the significance of this factor. (See, for example, Pacific Gas and Electric Company, 1982a, p. IV-23.)

c. Awareness. Some surveys found that the participants tended to show a greater awareness of conservation either by owning efficient appliances or by already having a ZIP loan. However, this evidence is controverted by other survey results. (See, for example, Pacific Gas and Electric Company, 1982a, p. IV-23). There was no reference in the survey results to the role played by the oil crisis in increasing the participants' awareness.

d. Portability. As to *portability,* we suggested that one would expect less interest in insulation among tenants than among home owners. As it turned out, several of those who did not participate in the audit program said that they felt that insulation is the landlord's responsibility (Pacific Gas and Electric Company, 1982a,b, p. III-6). However, among the participants, the ownership or rental of the dwelling did not seem to be important. But it (ownership or rental) was relevant for the do-it-yourself work: There was higher interest among renters in do-it-yourself projects than there was among home owners (See Pacific Gas and Electric Company, 1982a, IV-23.)

A little thought suggests that this should not be unexpected. While in general the duration of occupancy may be shorter for renters than for home owners, the payback for insulation may still be short enough to make it worthwhile for renters to insulate, even though they may feel that insulation is the landlord's responsibility. Furthermore, by doing it themselves, the tenants can shorten the payback period even further. At any rate, even if the renters do not subscribe to the insulation program themselves, we cannot conclude that rented dwellings will be less insulated than owner-occupied dwellings. To the extent that uninsulated dwellings have higher energy bills, it is probably reasonable to expect the tenants to resist paying the same rent for uninsulated and insulated dwellings. If that is true, the landlord is bound to heed the message and insulate.

One may expect the portability of the conservation measure to affect differentially the demand for insulation among various age-groups: at least one would expect the older age-groups to show less interest in insulation because of its nonportability. On the other hand, the short payback period for insulation may make age a moot question. Pacific Gas and Electric's RCS follow-up found that the RCS participants are as likely to be 65 years of age and over as were the nonparticipants. The survey found no evidence that age plays an important role in determining the participation in conservation measures in general (Pacific Gas and Electric Company, 1982a, p. IV-20). Similarly Edison found that there was no significant difference in the age of the participants compared to the age of the nonparticipants (Southern California Edison, 1982, 4 XIII-4-6).

e. Interest Rate and Initial Cost. The evidence suggests that lower interest rates were a major attraction of programs that provided low-interest loans. The interest-free nature of ZIP was mentioned as the reason for joining the program by 65% of the respondents (Pacific Gas and Electric Company, 1982b, p. VI-5). The ZIP program also attracted disproportionately those who had been the least motivated to install conservation equipment and devices (Pacific Gas and Electric Company, 1982b, p. VII-6). Close to three-fifths of the participants said that they would not have insulated or adopted the conservation measures they adopted if they had not obtained them through an interest-free program. Another one-fifth said that they would have delayed the adoption of these measures for at least another year and possibly longer.

In comparison with the initial cost of the conservation measure, the interest-rate factor seems to be the dominant one. In fact, the answers from the follow-up studies do not suggest that the initial cost, as well as factors affecting the initial cost, played a key role in the household demand for conservation measures, including insulation.

It is also of more than passing interest to note, for example, that the majority of the participants in PG&E's ZIP program expressed preference for an interest-free loan over a 50% rebate of the total cost of the conservation measures. Among nonparticipants the opposite was true. However, almost as many among the nonparticipants were undecided (Pacific Gas and Electric Company, 1982a, p. VI-10).

It is also interesting that about one-fifth of those who participated in PG&E's programs were not even aware they were entitled to a tax credit for the expenditures incurred for the conservation measures they adopted. Moreover, only 69% of those who participated in the ZIP program claimed a tax credit (Pacific Gas and Electric Company, 1982b, p. VII-20).

2. Summary

Table IV summarizes the results that emerge from our examination of the role played in 1980, and possibly 1979, by the theoretical determinants identified in Section II.A. These results show (1) that the key determinants include energy price, energy use per household, and interest rate; (2) that the secondary variables include initial cost, intangible cost, awareness as reflected in ownership of efficient appliances, nonportability of insulation, and type of dwelling; and (3) that factors about which the surveys are silent include direct effect of income on insulation, effect of awareness due to the oil crisis, effect of awareness due to utility initiatives and programs, and harmful side effects of insulation.

Of the four factors listed in the third group, the classification of the last two comes as no surprise. The harmful side effect of insulation has probably been overplayed. The utility program and initiative are part of

TABLE IV

A Summary of the Significance of the Theoretical Determinants of Insulation Demand as Viewed by the Survey Respondents

Theoretical variable	Significance according to respondents
Factors increasing insulation demand	
Energy price	Important
Energy use per household	Important
Factors decreasing insulation demand	
Interest rate	Important
Initial cost	Secondary
Income	No reference
Intangible cost	Secondary
Factors increasing the awareness of need for insulation	
The oil crisis	No reference
Utility initiative	No reference
Ownership of efficient appliances	Secondary
Harmful side effects of insulation	No reference
Other factors affecting insulation demand	
Portability	Secondary
Type of dwelling	Secondary

a continuum, one medium out of many, that have been enhancing the consumer's awareness on a continuing basis. To the extent that the utility program exerts a separate impact, it does so through a unique feature that a particular program has. An outstanding example is the interest-free aspect of the ZIP program. Its impact is reflected in the effect of the lower interest rate on insulation demand. Once this effect is accounted for, the existence of the utility program as an awareness medium merges with the other media, and it probably becomes impossible to differentiate it from the rest.

The classification of the first two variables in group three, while surprising, is explainable. Income, being a key factor in any household's economic decisions, may be expected to exert a direct effect of its own on the decision to insulate separate from its indirect effect through energy use. The surveys do list income as a major characteristic of households that demand conservation measures, but it is probably too much to expect the surveys to differentiate between the direct and indirect components. They were not pitched for that level of analysis. We will group this variable with the key variables in our model specification even though the survey results lack analysis of its direct and indirect effects.

Similar remarks apply to the omission of the awareness role played by the oil crisis from the respondent's answers. The respondents were asked specific questions about their actions in 1979 or 1980, six or seven years after the period when the oil crisis occurred. There was no occasion for the respondents, as far as one can see from the questionnaires, to go back to 1973 and probe how it all started. The survey compilers did not undertake such an analysis either. Since here, as in the case of income, there is a strong *a priori* reason to believe that the oil crisis played a catalytic role by enhancing the consumer's sensitivity to changes in energy prices, we will include it in the model.

Two more points before we wrap up this section:

1. In addition to the results just reported, the survey results confirm that consumers do not make the adjustment to a higher insulation level immediately, but rather go through a period of varying length.
2. The results do not suggest that households make a lifetime cost calculation for insulation or other conservation measures, although several of these factors do enter their decisions on whether to adopt a certain conservation measure.

C. Model Specification

In line with our discussion of the adjustment process in Section II.A, we postulate that the insulation demand follows a partial adjustment process, i.e.,

$$\Delta I_{\text{IN}_t} = s(I^*_{\text{IN}_t} - I_{\text{IN}_{t-1}}) + u_t, \tag{1}$$

where I_{IN_t} is an index of insulation, $0 < s \le 1$ is the speed of adjustment parameter, and u_t is a random term. We assume that the target insulation level $I^*_{\text{IN}_t}$ is determined according to

$$I^*_{\text{IN}_t} = a_0 + a_1 C_{\text{HH}_t} + a_{21} P_{1t} + a_{22} P_{2t} + a_3 r_t + a_4 \text{YPC}_t, \tag{2}$$

where C_{HH_t} is energy consumption per household, P_{1t} real price of energy 1969–1981, P_{2t} real price of energy 1974–1981, r_t real rate of interest, and YPC real income per household. We expect that $a_1 > 0$, $a_{21} > 0$, $a_{22} > 0$; $a_3 < 0$, $a_4 < 0$.

Substituting (2) in (1), we obtain the generic form of the model:

$$I_{\text{IN}_t} = s(a_0 + a_1 C_{\text{HH}_t} + a_{21}P_{1t} + a_{22}P_{2t} + a_3 r_t + a_4 \text{YPC}_t)$$
$$+ (1 - s)I_{\text{IN}_{t-1}} + u_t. \tag{3}$$

Two observations are in order.

1. Following Goldberger (1972), we could have specified the unobservable variable to be a stochastic function having a normally distributed random term with zero mean, constant variance, and zero covariance with u_t in (3). The interpretation of (1) will be slightly different. Now I_{IN_t} will lag behind the *expected value* of $I^*_{\text{IN}_t}$. Similarly, under the stochastic specification of (2), the random term in (3) will be a composite term. But our estimates will remain unchanged under either specification.
2. Instead of specifying (1) and (2) in linear form, we could have specified both as linear in the logarithm. Equation (3) would have been linear in the logarithm as well. Both the linear and the double logarithmic forms are subsets of a larger set of functional specification defined by the Box–Cox transformation. Since there is no *a priori* reason for expecting (3) to take either form, linear or loglinear, we will not constrain the functional specification in advance. Instead, we will estimate with a Box–Cox transformation and let the data determine the functional form of our model. We take up this subject in Section IV.A.

The derivation of the insulation index I_{IN_t} and its operational definition are explained in detail in Section III. The operational definition of the remaining variables in (3) is taken up in IV.C.

Equation (1) implies that I_{IN_t} lags geometrically behind $I^*_{\text{IN}_t}$. Equivalently, this implies that I_{IN_t} is a function of the expected level of energy consumption per household, energy price, interest rate, and income per household, where the expected values are defined in terms of the geometrically distributed lag of these variables. [The same is true of the Box–Cox transform of (3).] This is consistent with the view that a household's expectations are influenced by its experience with prices, interest rates, etc., rather than just by current values. It also implies that when energy prices, household income, etc., increase, their strongest impact on insulation demand occurs during the period of the increase, but that the impact continues in subsequent periods at constantly decreasing rates.

Two more observations.

1. Equation (3) includes a variable P_{2t} to capture the impact of increased awareness ushered in by the oil crisis. Since the focus during that period was on the steep escalation of oil prices, it is reasonable to assume that the effects operated through the price level by strengthening the relationship between energy price and insulation demand. Hence, we may expect an increase in energy price to have resulted in a larger increase in the demand for insulation in the post- than in the pre-oil-crisis period. This is the reason for including P_{2t} (in addition to P_{1t}) in the equation by defining it to be zero in 1969–1973 and P_t from 1974 on.

   One may be tempted to argue that in a similar manner we should expect the utility's programs to have resulted in a further strengthening of the relationship between insulation demand and energy price in the post-1978 period. But as we pointed out, the utility's efforts are not the only source of enhanced awareness. There has been a proliferation of agencies, advertisements, television shows, and radio discussions that served to increase the consumer's awareness of the insulation option. This diffusion of awareness through various channels tends to blur the demarcation within the post-oil-crisis era and makes it difficult to argue that a well-defined era was ushered in with the introduction of utility programs in the same way that the quadrupling of oil prices in 1979 had ushered in a new era. (This means also that, in practice, it will not be possible to sort out in any statistically meaningful sense the separate contribution of the utility's efforts to the increased responsiveness, if any, of insulation demand to energy prices.)

2. It is important not to lose track of the fact that the demand for insulation is a demand for one alternative to electricity usage, albeit an important alternative. The demand relationship specified in (3) should be viewed as only one in a system of demand relationships which include the demand for more efficient appliances[11] (as well as the demand for other energy sources of space heating and air conditioning, such as solar heating, wood burning, and solar air conditioning). We hope to turn to the development and estimation of a model of the demand for efficient appliances in the future.

---

[11] Time series of indices for the efficiency level of the stock of individual appliances have been developed as part of the work reported in Khazzoom (1986).

III. THE CONSTRUCTION OF THE INSULATION INDEX

In this section we shall describe the derivation of our insulation index. During most of the 1969–1980 sample period, insulation was about the only factor known to have affected the residential thermal efficiency in SMUD's service area.

Briefly, to develop our index, we first constructed estimates of the distribution of dwellings by insulation, measured by resistance levels R-7, R-11, etc. Next we summarized these distributions by constructing indices of the insulation level of each dwelling type with a given heating or cooling mode. To do that we ran through an engineering model every dwelling type with a given heating or cooling mode and derived for each year engineering estimates of the energy requirements for resistance heating, heat-pump heating, heat-pump cooling, and air conditioning for that dwelling. We then translated these energy requirements into summary measures (indices) of the insulation by dwelling type. In the final step, we calculated indices for aggregates for various groupings of dwellings.

Over time, changes take place in the square footage of dwellings, in the number of dwellings of each type (single-family, duplex, etc.), and in the heating or cooling mode. In order for the index of aggregates to be meaningful, the goal should be to define a procedure that can calculate the insulation index while holding constant such variables as the square footage of the dwelling and the number of dwellings of each type with each heating or cooling mode. It is also desirable that the index posseses two more attributes: (1) It should be unbiased for any grouping of dwellings for which we chose to calculate the index (the grouping may be all single-family and duplex homes or all resistance- and gas-heated multiple-family homes, and so on); (2) our index should consist of a single measure; that is, we would like to be able to represent the insulation level by one index rather than, say, by two indices (which may yield divergent results over time).

A. Observations on the Data Compiled for the Insulation Index

To derive our insulation index, we needed first to compile two sets of basic data. These include the following:

1. Distribution of customers by dwelling type (single-family, multiple-family, duplex, and mobile homes), by heating and cooling mode, and by energy source (gas versus electricity); and

2. distribution by level of insulation of electrically heated homes and by type of heating (resistance heat and heat pump) and cooling (air conditioning and heat-pump cooling).

We compiled these data from separate monthly records in SMUD's files, and we collected similar data for gas-heated homes. We collected separate data for electrically heated homes because of the tremendous acceleration in the number of customers with electrically heated homes since 1969 and because SMUD's electricity price (SMUD generates

TABLE V

Distribution of Insulation Level of Ceiling (C) and Wall (W) of Electrically Heated Single-Family Homes in SMUD's Service Area Using Resistance Heat [a]

		Number of units distributed by R level										
Year	Annual average	C 7	W 0	C 7	W 7	C 13	W 7	C 13	W 11	C 19	W 11	C 22	W 11
1960	700	700										
1961	810	700	110									
1962	938	700	238									
1963	1,085	700	238	147								
1964	1,256	700	238	318								
1965	1,454	700	238	516								
1966	1,684	700	238	746								
1967	1,949	700	238	1,011								
1968	2,258	700	238	1,320								
1969	2,614	700	238	1,676								
1970	2,790	700	238	1,764	88							
1971	3,262	700	238	2,000	324							
1972	4,018	700	238	2,378	702							
1973	5,005	700	238	2,872	1,195							
1974	5,869	700	238	3,304	1,627							
1975	6,541	700	238	3,304	1,627	672						
1976	6,983	700	238	3,304	1,627	1,114						
1977	8,815	527	175	3,290	1,620	3,182	21					
1978	11,515	314	98	3,273	1,611	6,172	47					
1979	14,367	173	47	3,262	1,605	9,216	64					
1980	15,161	68	NA	3,253	1,601	10,162	77					
1981	15,270	NA	NA	3,243	1,596	10,339	92					

[a] Sources: Distribution of R levels: Compiled on the basis of records of Division B, SMUD's Conservation Department. Retrofit estimates for 1977–1981 were provided by the Planning Division of SMUD's Conservation Department.

50% of its output from hydropower and 50% from nuclear power) has not been and is not likely to become a simple function of the behavior of gas prices in the future.

In estimating our model, we used broader aggregates than the data we collected. We made estimates for all electrically heated homes (winter and separately summer) and all gas-heated homes during the summer. Data on the aggregate sales to electrically and gas-heated homes, as well as the average number of customers with gas and electric heat, are taken from SMUD's records of rates 14, 15, 17, 18, which cover the electrically and gas-heated homes. These were given in Tables II and III.

Table V, compiled for resistance-heated single-family homes, illustrates the type of data we compiled on the distribution of insulation level for electrically heated homes. Table VI illustrates the case for gas-heated mobile homes. We prepared eight additional tables: (1) single-family with heat pump, (2) duplex with resistance, (3) duplex with heat pump, (6) single-family with gas heat, (7) duplex with gas heat, and (8) multiple-family with gas heat. For details, see Khazzoom (1986).

To see how we calculated the distribution of insulation for the stock of dwellings, consider the stock of a dwelling type with a given heating mode, e.g., a single-family dwelling with resistance heating and a ceiling insulated at a particular level of resistance, call it R_k. We added to this stock the newly constructed homes as well as the number of homes retrofitted to this insulation level during the year. We subtracted the number of homes whose initial ceiling insulation level was R_k but was upgraded to a higher resistance level during the same period.

Where $S_{st}^{R_k}$ is the stock of single-family homes with insulation level R_k, at the beginning of t; $I_{st}^{R_k}$ is new construction of single-family homes with insulation level R_k; and $RF_{st_{R_j}}^{R_i}$ is the number of single-family homes retrofitted to R_i from an existing insulation level R_j, we have

$$S_{st}^{R_k} = S_{st-1}^{R_k} + I_{st}^{R_k} + \sum_{i \neq j} RF_{st_{R_j}}^{R_i} \qquad (4)$$

with the following contribution of retrofit to the stock

$$RF_{sR_j}^{R_j} \begin{cases} > 0 & \text{when} \quad i = k, \\ < 0 & \text{when} \quad j = k, \\ = 0 & \text{otherwise.} \end{cases} \qquad (5)$$

TABLE VI

Distribution of Insulation Level of Ceiling (C), Floor (F), and Wall (W) of Gas-Heated Mobile Homes in SMUD's Service Area [a]

Year	Annual average	Number of units distributed by R level			
		C F W 0 0 0	C F W 8.2 5.5 5.5	C F W 16 10 8	C F W 19 11 11
1960	2,821	2,821			
1961	2,986	2,986			
1962	3,107	3,107			
1963	3,254	3,254			
1964	3,405	3,405			
1965	3,521	3,521			
1966	3,632	3,632			
1967	3,731	3,731			
1968	3,934	3,934			
1969	4,139	4,139			
1970	4,333	4,333			
1971	4,912	4,743	169		
1972	5,751	4,743	1,008		
1973	6,862	4,743	2,119		
1974	7,800	4,743	2,119	938	
1975	8,448	4,743	2,119	1,586	
1976	8,905	4,743	2,119	2,043	
1977	9,686	4,743	2,119	2,043	781
1978	10,347	4,743	2,119	2,043	1,442
1979	11,011	4,743	2,119	2,043	2,106
1980	11,614	4,743	2,119	2,043	2,709

[a] Sources: Distribution by Rs: 1960–1970: No regulations of insulation level existed during this period. The Rs in this table were constrained at zero on the basis of estimates by personnel from the industry and the Department of Housing and Community Development. 1971–1973: California Administrative Code, Title 25, Chapter 3, September 15, 1971. We prorated the additions to mobile homes in 1971 to reflect the timing of Title 25. 1974–1976: California Administrative Code, Title 25, Chapter 3, January 1, 1974. 1977–1980: June 15, 1976, the National Mobile Home Construction Safety Standards Act of August 22, 1974, went into effect. It specified the U factor for the envelope of the mobile home. Based on conversation with HUD officials, we have used the FHA's insulation requirements for 1977 on the single-family home instead and added the floor insulation level of R-11.

A similar expression as in (4) and (5) applies to the stock of any other residence type (multifamily, duplex, and mobile homes). Since retrofitting involves homes built in an earlier period but which have a square footage different from that of homes built in later years, the implementation of (4) requires us to keep a record not only of the distribution of insulation level (ceiling or walls or floor), but also of the distribution of dwelling size (square footage) for every combination of ceiling, wall, and floor insulation level.

We note limitations in our stock estimates.

1. In estimating the insulation level of homes of various types, we treated all the increase in the number of customers (e.g., the increase in single-family customers with electric heat) as representative of new home construction and applied to these the insulation levels known to have been applicable to newly constructed homes during that year. Some of the electrically heated dwellings may, in fact, have been added as a result of a switch from gas to electric heat. It is conceivable, although very unlikely, that the switch was made with a retrofitting but to a lower insulation level than that of newly built homes with electric heat. The same may be true of the net annual additions to gas-heated homes, some of which may have switched from electric heat. There will be errors due to this source, but they are unlikely to be large. By all counts, the switch from electric heat to gas heat has been negligible during the sample period.

2. Equation (4) does not allow for the deterioration of insulation level over time.

3. We have no records of home retrofitting to higher insulation level prior to 1977. As a result, (4) reflects the retrofitting of homes from 1977 on.

The effect of the second limitation is the exaggeration of the insulation level of the stock of dwellings. The effect of the third is the underestimation of the insulation level, at least prior to 1977. These two opposing tendencies reduce the bias in our insulation estimates. Unfortunately, we have no *a priori* information about the direction of the net bias or the way it changed from one year to the next.

B. Derivation of the Insulation Index for the Micro Unit

For the purpose of estimating our model, we need an index of the insulation level for all electrically heated homes and all gas-heated homes. The derivation of the index for these aggregates can be done without deriving first an index for each component—namely, single-family dwellings with resistance heat, single-family dwellings with heat pumps, duplexes with resistance heat, duplexes with heat pumps, etc.—and then aggregating over components. However, the derivation of the index for individual components should help to clarify the basic ideas.

We used the CAL PASS program of the Berkeley Solar Group to generate the engineering data necessary for calculating our indices. CAL PASS is an engineering model which calculates the energy required for homes of various types, sizes, structures, insulation levels, window glazing levels, etc.

To illustrate the derivation of the insulation index for a microunit, let us focus on single-family homes with resistance heat. Turn to Table V. Divide the number of homes listed under each insulation level by the total number of homes (shown in column 1) during that year to get the share of each insulation level in the stock of resistance-heated single-family homes in that year. So, for example, in 1960, 100% of the homes had a ceiling (C) insulation level of R-7 and wall (W) insulation level of R-0. In 1961, 86% (700 ÷ 810) of the homes had R-7 and R-0 for C and W, respectively, and 14% (110 ÷ 810) had R-7 and R-7 for C and W, respectively. In a similar manner, we found that in 1962, 75% (700 ÷ 938) and 25% (238 ÷ 938) had R-7 and R-0 and R-7 and R-7 for C and W, respectively.

Conceptually, we may view the year-to-year change in these shares as a change in the distribution of the insulation level of a synthetic home whose wall and ceiling insulation is upgraded annually but without changing the area from its 1960 level and while maintaining normal weather conditions. An increase in the share of higher insulation levels in the composition of this synthetic home leads to a reduction in the electricity required for heating and cooling. This relative reduction in energy requirement (which represents an increase in the thermal efficiency of the dwelling) is used in our system to define the increase in the insulation index. To see this, turn to column 1 of the panel marked winter in Table VII.

The engineering estimates (by CAL PASS) show that in 1960, 26,140 kW h were required to heat a 1250-ft^2 home (this is the actual average area of a 1960 single-family electrically heated home in SMUD's service area) under normal weather conditions. As a result of the upgrading of insulation, the electricity required in 1961 and 1962 to maintain the same level of comfort dropped to 24,305 and 22,712 kW h, respectively. In our system, we fix the insulation index in 1960 at 1.0 and define the reduction in the electricity requirement from 1960 to 1961 and from 1960 to 1962 as an increase in the insulation index from 1.0 in 1960 to 1.075 in 1961 ($26,140 \div 24,305$) and to 1.15 in 1962 ($26,140 \div 22,712$). That is, where E_{ij} is the energy required under normal weather conditions by a dwelling type i for end use j[12], we define

$$I_{\text{IN}_{ij}} = \frac{E_{ij_0}}{E_{ij_t}}. \tag{6}$$

Column 3 in Table VII lists $I_{\text{IN}_{ij}}$ for 1960–1981 for resistance-heated single-family dwellings.

In practice, what we did (which amounts to the same thing as what we just described) was to aggregate for each year the kilowatt hours required to resistance-heat the single-family homes listed in Table V, with area and all other home characteristics held as unchanged from their 1960 level. We then divided the total kilowatt hours of each year by the total number of single-family homes in that year, shown in column 1, Table V, and recorded the result (E_{ij_t}) in column 1, Table VII, under heating.

We will find the following alternative way of thinking about the insulation index to be useful in that it helps us to link our index of the microunits with our index of the aggregates. To take into account the changes in dwelling square footage and in the number of dwellings, let E_{ij}^a be the energy required under normal weather conditions by a dwelling type i, *with changing area* and for end use j, and N_{ij} be the number of dwellings of type i having end use j and then define an area index

$$I_{\text{A}_{ij_t}} = E_{ij_t}^a / E_{ij_t} \tag{7}$$

and a number-of-dwelling index

$$I_{\text{N}_{ij_t}} = N_{ij_t} / N_{ij_0}, \tag{8}$$

---

[12] Strictly, we should have attached a triple subscript to E, E_{hij}, where the subscript h denotes the heating mode (gas heat, resistance heat, or heat-pump heat), i denotes the type of dwelling, and j denotes the end use (space heating or air conditioning). To avoid the excessive use of subscripts, we did not include the h, since in our estimation the only aggregation we use is one that lumps all electrically heated homes on one side and all gas-heated homes on the other. This is also the context in which we use our indices of the aggregates.

TABLE VII

Index of Insulation of Electrically Heated Single-Family Homes in SMUD's Service Area Using Resistance Heat

Year	kW h required per dwelling unit with improved insulation and constant area of 1250 ft^2 (1)	kW h required per dwelling unit with improved insulation but with changing area (2)	Index of insulation $I_{IN_{ij}} = $ row 1 col. 1: row i col. 1 (3)	Index of area $I_{A_{ij}} = $ row i, col. 2 row i, col. 1 (4)	Index of dwelling number $I_{N_{ij}} = $ row i col. 2 Table V ÷ row 1 col. 2 Table V (5)	Reciprocal of value index $I_{TOT_{ij}} = $ (3) ÷ [(4) · (5)] (6)	Total energy requirement $D_{ij} = $ row i col. 2 of this table · row i, col. 1 of Table V (7)	Saturation (8)
				Winter				
1960	26,140	26,140	1.000	1.000	1.000	1.000	18,297,832	1.000
1961	24,305	24,326	1.075	1.001	1.157	0.929	19,703,917	1.000
1962	22,712	22,773	1.151	1.003	1.340	0.857	21,360,829	1.000
1963	21,230	21,341	1.231	1.005	1.550	0.790	23,155,366	1.000
1964	19,943	20,118	1.311	1.009	1.794	0.724	25,268,742	1.000
1965	18,830	19,082	1.388	1.013	2.077	0.659	27,745,627	1.000
1966	17,867	18,205	1.463	1.019	2.406	0.597	30,657,523	1.000
1967	17,038	17,472	1.534	1.025	2.784	0.537	34,052,365	1.000
1968	16,318	16,854	1.602	1.033	3.226	0.481	38,057,414	1.000
1969	15,699	16,346	1.665	1.041	3.734	0.428	42,728,597	1.000
1970	15,427	16,127	1.694	1.045	3.986	0.407	44,993,624	1.000
1971	14,841	15,663	1.761	1.055	4.660	0.358	51,091,316	1.000
1972	14,190	15,157	1.842	1.068	5.740	0.300	60,902,482	1.000
1973	13,636	14,737	1.917	1.081	7.150	0.248	73,760,645	1.000
1974	13,304	14,494	1.965	1.089	8.384	0.215	85,067,198	1.000
1975	13,024	14,182	2.007	1.089	9.344	0.197	92,766,944	1.000
1976	12,869	14,013	2.031	1.089	9.976	0.187	97,850,786	1.000
1977	12,069	13,158	2.166	1.090	12.593	0.158	115,987,975	1.000
1978	11,414	12,468	2.290	1.092	16.450	0.127	143,572,871	1.000
1979	11,086	12,134	2.358	1.095	20.524	0.105	174,331,917	1.000
1980	10,944	11,906	2.388	1.095	21.659	0.101	181,717,310	1.000
1981	10,871	11,909	2.405	1.095	21.814	0.101	181,850,686	1.000

TABLE VII (*Continued*)

Year	kW h required per dwelling unit with improved insulation and constant area of 1250 ft² (1)	kW h required per dwelling unit with improved insulation but with changing area (2)	Index of insulation $I_{IN_{ij}}$ = row 1 col. 1 : row i col. 1 (3)	Index of area $I_{A_{ij}}$ = row i, col. 2 row i, col. 1 (4)	Index of dwelling number $I_{N_{ij}}$ = row i col. 2 Table V ÷ row 1 col. 2 Table V (5)	Reciprocal of value index $I_{TOT_{ij}}$ = (3) ÷ [(4) · (5)] (6)	Total energy requirement D_{ij} = row i col. 2 of this table · row i, col. 1 of Table V (7)	Saturation (8)
				Summer				
1960	7,179	7,179	1.000	1.000	1.000	1.000	5,025,255	0.147
1961	6,792	6,797	1.057	1.001	1.350	0.783	6,421,478	0.172
1962	6,456	6,471	1.112	1.002	1.825	0.608	8,266,228	0.201
1963	6,113	6,139	1.174	1.004	2.462	0.475	10,581,283	0.234
1964	5,815	5,855	1.235	1.007	3.327	0.369	13,634,796	0.273
1965	5,557	5,614	1.292	1.010	4.494	0.285	17,662,031	0.319
1966	5,334	5,410	1.346	1.014	6.076	0.218	23,009,130	0.372
1967	5,142	5,240	1.396	1.019	8.204	0.167	30,087,807	0.434
1968	4,975	5,096	1.443	1.024	10.260	0.137	36,594,794	0.468
1969	4,832	4,977	1.486	1.030	12.825	0.112	44,678,324	0.506
1970	4,769	4,926	1.505	1.033	14.779	0.099	50,961,663	0.546
1971	4,635	4,819	1.549	1.040	18.393	0.081	62,043,825	0.581
1972	4,485	4,702	1.601	1.048	24.117	0.063	79,384,062	0.619
1973	4,358	4,605	1.647	1.057	31.032	0.050	100,030,174	0.639
1974	4,282	4,548	1.677	1.062	37.567	0.042	119,609,667	0.660
1975	4,208	4,482	1.706	1.065	43.264	0.037	135,745,767	0.682
1976	4,167	4,446	1.723	1.067	47.644	0.034	148,289,928	0.703
1977	3,963	4,260	1.812	1.075	62.050	0.027	185,051,564	0.726
1978	3,795	4,111	1.892	1.083	83.266	0.021	239,599,741	0.746
1979	3,710	4,039	1.935	1.089	106.732	0.017	301,788,398	0.766
1980	3,674	4,006	1.954	1.091	115.704	0.015	324,495,304	0.787
1981	3,657	3,989	1.963	1.091	119.719	0.015	334,328,939	0.808

with $I_{A_{ij_0}} = I_{N_{ij_0}} = 1.0$. Our insulation index $I_{IN_{ij}}$ can then be seen to be one component in a multiplicative decomposition of the reciprocal of a value index, well known from index-number theory, which relates the total energy requirement estimated by CAL PASS for an initial period, call it D_{ij_0}, to the total requirement estimated by CAL PASS for a subsequent period D_{ij_t}. Denoting this reciprocal of the value index by $I_{TOT_{ijt}}$, we have

$$I_{TOT_{ijt}} = D_{ij_0}/D_{ij_t} = (E_{ij_0} \times I_{A_{ij_0}} \times N_{ij_0})/(E_{ij_t} \times I_{A_{ij_t}} \times N_{ij_t}) \qquad (9)$$

$$= I_{IN_{ijt}} \times (I_{A_{ijt}})^{-1} \times (I_{N_{ijt}})^{-1}, \qquad (10)$$

which shows $I_{IN_{ijt}}$ as one ingredient in a multiplicative decomposition of D_{ij_0}/D_{ij_t}. The D_{ij}'s for resistance-heated single-family dwellings are shown in column 7 of Table VII for 1960–1981; $I_{TOT_{ij}}$ is shown in column 6 of the same table for 1960–1981.

The reciprocal value index D_{ij_0}/D_{ij_t} in (9) serves as the basis for our aggregation over groups of dwellings with the same end use or with different end uses. In fact, the structure of our index of the aggregates [see Eqs. (16) and (17) later in this chapter) is identical to the structure in (10).

A brief explanation of the idea behind (7) is in order. An increase in the area of a dwelling increases the electricity required to heat and to cool that dwelling, other things being equal. But the increase in the electricity requirement is not proportional to the increase in the area. It depends on the type of home and its insulation level, as well as on its other characteristics. Hence, to derive an index of the effect of a change in the square footage on the electricity requirement we used the same idea of a synthetic home whose insulation is being upgraded, except that we now allow the area of this home to increase (or to decrease) in the same way as it has changed historically in SMUD's service area. We reran CAL PASS as we did before, but with added information on the changing area, to calculate the E^a_{ij}, i.e., the energy required to heat (or to cool) the dwelling with the larger or smaller square footage. The results of calculating E^a_{ij} for resistance-heated single-family dwellings are shown in column 2 of Table VII. As expected, a comparison of the heating requirements in columns 1 and 2 shows that the electricity requirement has now increased. Nothing has changed in the transition from column 1 to column 2 except the area. Hence we define the area index I_A as the ratio of column 2 to column 1. This is what (7) shows. The results are listed in column 4 of Table VII.

To calculate the index for the number of homes in (8), turn to Table V. You will find that the number of single-family dwellings with resistance heat increased from 700 in 1960 to 15,270 in 1981. Substituting these data in (8), we find $I_{N_{ij_t}} = 21.814$ in 1981, which is shown in column 5 of Table VII. The index is calculated in a similar manner for the intervening years. Also, $I_{A_{ij}}$ and $I_{N_{ij}}$ for electrically heated single-family dwellings using heat pumps or for gas-heated single-family dwellings are calculated in the same manner. The same is true of all other types of dwellings.

We do not make use in this study of the ancillary area and number of dwellings indices I_A and I_N. They are a by-product of our decomposition of the reciprocal value index D_{ij_0}/D_{ij_t}. But they become critical when we are interested in going beyond the demand for insulation and inquiring how the resulting higher insulation level affects the demand for electricity or gas when both the area and the number of dwellings change at the same time. Then the two indices I_A and I_N provide the required link for making the transition from insulation demand to energy demand. These two indices will appear also with the insulation index of the aggregates we derive subsequently.

We reiterate that the insulation indices we derive in this study are based strictly on an engineering concept, which means that in calculating the index, the level of heating comfort and similarly all other home characteristics remain unchanged from their 1960 level. All that is allowed to change in the interim is the insulation level. Hence changes in energy requirement reflect solely the effect of the change in insulation.

The summer index of insulation is derived in exactly the same manner. It is shown in column 3 of Table VII under the summer panel. As we might expect, this index shows a different insulation level from the index reported under the winter panel. There are two reasons: First, the proportion of dwellings with resistance heat that have air conditioning increased over time. (Note the air conditioning saturation—column 8, Table VII under the summer panel.) This is also reflected in $I_{IN_{ij}}$, which increased six times as fast in the summer as in the winter for resistance-heated dwellings (see summer and winter panels, column 5). Second, even with the number of homes remaining constant, a change in the insulation level has a different effect on energy requirements in the summer and winter.

As it is, column 3 of Table VII shows that for single-family dwellings, the insulation index stood at 2.405 for heating and 1.963 for cooling in 1981. We could have combined the index of insulation for the summer and winter into one index for the whole year and estimated an annual model without differentiating between summer and winter. We chose not to do that, since, as we said, the same level of R insulation has a different impact on energy demand in the summer and the winter, and it is through its impact on energy demand that, in the final analysis, insulation is judged.

C. Aggregation over Dwellings and End Uses

Since we are interested in estimating the model for electrically heated and gas-heated dwellings, we aggregate first over all electrically heated dwellings and only for those uses that are affected by insulation—space heating and separate air conditioning. We do the same for gas-heated homes. The results are shown in Tables VIII and IX. We could have aggregated over a different grouping to derive year-round demand for insulation by all electrically heated homes. Alternatively, we could have aggregated heating and cooling demand over a smaller subset of dwellings, such as single-family dwellings, rather than all electrically heated dwellings, and so on. Indices for levels of aggregation of heating and cooling, which are different from those we used in our estimation, are not reported here. See Khazzoom (1986), Tables B-21 to B-25, B-27 to B-29, and B-31.

We define the aggregate counterpart of (9) as

$$I_{\text{TOT}} = \sum_{ij} D_{ij_0} \Big/ \sum_{ij} D_{ij_t}, \tag{11}$$

where we aggregate over the subset of dwellings using the end uses in which we are interested. Equation (11) can be simplified by noting that

$$D_{ij_t} = D_{ij_0}/I_{\text{TOT}_{ij_t}} \tag{12}$$

$$= D_{ij_0}/[(I_{\text{IN}_{it_t}}) \times (I_{A_{ij_t}})^{-1} \times (I_{N_{ij_t}})^{-1}] \tag{13}$$

from (9) and (10). Making the substitution from (12) into (11) and dividing the numerator and denominator by $\Sigma_{i,j}\, D_{ij_0}$, we have

$$I_{\text{TOT}_t} = \frac{\Sigma_{i,j}\, D_{ij_0}}{\Sigma_{i,j}\, D_{ij_t}} = \frac{\Sigma_{i,j}\, D_{ij_0}}{\Sigma_{i,j}\, (D_{ij_0}/I_{\text{TOT}_{ij_t}})}$$

$$= \left[\sum_{i,j} s_{ij_0}(I_{\text{TOT}_{ij_t}})^{-1} \right]^{-1} \tag{14}$$

$$= \sum_{i,j} s_{ij_0}(I_{\text{IN}_{ij_t}})^{-1} \times I_{\text{A}_{ij_t}} \times (I_{\text{N}_{ij_t}})^{-1}, \tag{15}$$

where we made use of (13) in deriving (15). (Recall that the D_{ij}'s are engineering requirements derived from CAL PASS, *not* behavioral demand.)

Equation (14) gives I_{TOT} (the reciprocal value index of the aggregate) as the reciprocal of the weighted average of the corresponding microindices, where the weights are the base-period share of energy requirements of each home type with a given end use. A similar interpretation applies to (15) except that here we show the microinsulation indices explicitly in the expression. When $I_{\text{A}_{ij_t}} \equiv I_{\text{N}_{ij_t}} \equiv 1$, (15) reduces to the weighted insulation index of the aggregate. It is straightforward to demonstrate that the index in (15) is unbiased in the sense that when it is applied to the aggregate energy requirements in a base period [to derive the effect of increased (or decreased) insulation], it predicts a total requirement for t which is identical to the aggregate of the individual requirements predicted for t by the microindices.

D. A Multiplicative Decomposition and a Derivation of the Weighted Insulation Index for the Aggregate

An alternative and perhaps a more revealing way of writing the decomposition of (11) is shown in the equations that follow, where for the sake of simplicity we suppressed the subscripts i, j:

$$I_{\text{TOT}_t} = \underbrace{\frac{\sum E_0(N_0 I_{\text{A}_0})}{\sum E_t(N_0 I_{\text{A}_0})}}_{(I_{\text{IN}_t})} \times \underbrace{\frac{\sum(E_t N_t) I_{\text{A}_0}}{\sum(E_t N_t) I_{\text{A}_t}}}_{(I_{\text{A}_t})^{-1}} \times \underbrace{\frac{\sum(E_0 I_{\text{A}_0})(E_t/E_0) N_0}{\sum(E_0 I_{\text{A}_0})(E_t/E_0) N_t}}_{(I_{\text{N}_t})^{-1}} \tag{16}$$

$$= \underbrace{\frac{\sum E_0(N_t I_{\text{A}_t})}{\sum E_t(N_t I_{\text{A}_t})}}_{(I_{\text{IN}_t})} \times \underbrace{\frac{\sum[I_{\text{A}_0}(E_0 N_0)]}{\sum[I_{\text{A}_t}(E_0 N_0)]}}_{(I_{\text{A}_t})^{-1}} \times \underbrace{\frac{\sum(E_t I_{\text{A}_t})(E_0/E_t) N_0}{\sum(E_t I_{\text{A}_t})(E_0/E_t) N_t}}_{(I_{\text{N}_t})^{-1}}. \tag{17}$$

Equation (16) can be shown to be equivalent to (11) by rewriting (15) as

$$I_{\text{TOT}_t} = \left[\sum_{ij} s_{ij_0}(I_{\text{IN}_{ij_t}})^{-1}\right]^{-1} \times \frac{\sum_{ij} s_{ij_0}(I_{\text{IN}_{ij_t}})^{-1}}{\sum_{ij} s_{ij_0}(I_{\text{IN}_{ij_t}})^{-1} \times I_{\text{A}_{ij_t}}}$$

$$\times \frac{\sum_{ij} s_{ij_0}(I_{\text{IN}_{ij_t}})^{-1} I_{\text{A}_{ij_t}}}{\sum_{ij} s_{ij_0}(I_{\text{IN}_{ij_t}})^{-1} I_{\text{A}_{ij_t}} \times I_{\text{IN}_{ij_t}}}. \tag{18}$$

TABLE VIII

Index of Insulation of All Electrically Heated Homes in SMUD's Service Area

Year	Equation (16)				Equation (17)							
	Index of insulation (1)	Index of area (2)	Adjusted index of the number of dwellings (3)	Index of total effect = $(1) \div [(2) \cdot (3)]$ (4)	Index of insulation (5)	Index of area (6)	Adjusted index of the number of dwellings (7)	Index of total effect = $(5) \div [(6) \cdot (7)]$ (8)	Fisher's ideal index of insulation = $[(1) \cdot (5)]^{1/2}$ (9)	Fisher's ideal index of area = $[(2) \cdot (6)]^{1/2}$ (10)	Fisher's ideal index of the number of dwellings = $[(3) \cdot (7)]^{1/2}$ (11)	Fisher's ideal index of total effect = $[(4) \cdot (8)]^{1/2}$ (12)
					Winter							
1960	1.000	1.000	1.000	1.000	1.000	1.000	1.000	1.000	1.000	1.000	1.000	1.000
1961	1.054	1.001	1.137	0.926	1.055	1.001	1.138	0.926	1.055	1.001	1.138	0.926
1962	1.107	1.002	1.297	0.852	1.109	1.002	1.299	0.852	1.108	1.002	1.298	0.852
1963	1.174	1.005	1.476	0.791	1.181	1.004	1.486	0.791	1.177	1.005	1.481	0.791
1964	1.239	1.009	1.683	0.730	1.253	1.007	1.705	0.730	1.246	1.008	1.694	0.730
1965	1.300	1.012	1.922	0.668	1.325	1.011	1.961	0.668	1.312	1.011	1.941	0.668
1966	1.358	1.017	2.199	0.607	1.396	1.015	2.264	0.607	1.377	1.016	2.231	0.607
1967	1.414	1.023	2.520	0.548	1.465	1.021	2.619	0.548	1.439	1.022	2.569	0.548
1968	1.466	1.030	2.898	0.491	1.532	1.027	3.039	0.491	1.498	1.029	2.968	0.491
1969	1.514	1.039	3.335	0.437	1.594	1.033	3.531	0.437	1.553	1.036	3.432	0.437
1970	1.544	1.046	3.722	0.397	1.640	1.038	3.985	0.397	1.591	1.042	3.851	0.397
1971	1.607	1.065	5.026	0.300	1.744	1.047	5.548	0.300	1.674	1.056	5.281	0.300
1972	1.664	1.083	6.748	0.228	1.828	1.057	7.596	0.228	1.744	1.070	7.160	0.228
1973	1.709	1.097	8.582	0.182	1.887	1.067	9.742	0.182	1.796	1.082	9.144	0.182
1974	1.741	1.106	10.156	0.155	1.930	1.073	11.604	0.155	1.833	1.090	10.856	0.155
1975	1.772	1.105	11.581	0.138	1.963	1.071	13.235	0.138	1.865	1.088	12.381	0.138
1976	1.800	1.104	12.981	0.126	1.986	1.069	14.788	0.126	1.891	1.087	13.855	0.126
1977	1.915	1.104	15.980	0.109	2.062	1.066	17.820	0.109	1.987	1.085	16.875	0.109
1978	2.022	1.104	19.976	0.092	2.138	1.061	21.966	0.092	2.079	1.082	20.947	0.092
1979	2.128	1.099	24.764	0.078	2.187	1.056	26.489	0.078	2.157	1.077	25.612	0.078
1980	2.228	1.080	28.765	0.072	2.206	1.051	29.261	0.072	2.217	1.065	29.012	0.072

Year												
1960	1.000	1.000	1.000	1.000	1.000	1.000	1.000	1.000	1.000	1.000	1.000	1.000
1961	1.021	1.000	1.094	0.933	1.022	1.000	1.096	0.933	1.022	1.000	1.095	0.933
1962	1.040	1.001	1.206	0.862	1.045	1.000	1.212	0.862	1.043	1.001	1.209	0.862
1963	1.066	1.001	1.342	0.793	1.080	1.001	1.360	0.793	1.073	1.001	1.351	0.793
1964	1.091	1.002	1.508	0.722	1.118	1.001	1.547	0.722	1.104	1.002	1.528	0.722
1965	1.114	1.004	1.717	0.647	1.160	1.002	1.791	0.647	1.137	1.003	1.753	0.647
1966	1.137	1.006	1.981	0.570	1.205	1.003	2.108	0.570	1.171	1.004	2.043	0.570
1967	1.159	1.009	2.321	0.495	1.254	1.003	2.527	0.495	1.205	1.006	2.422	0.495
1968	1.180	1.013	2.659	0.438	1.299	1.004	2.954	0.438	1.238	1.009	2.803	0.438
1969	1.199	1.018	3.059	0.385	1.343	1.005	3.470	0.385	1.269	1.011	3.258	0.385
1970	1.215	1.022	3.491	0.341	1.383	1.006	4.036	0.341	1.296	1.014	3.754	0.341
1971	1.250	1.032	4.686	0.258	1.468	1.006	5.644	0.258	1.355	1.019	5.143	0.258
1972	1.271	1.043	6.216	0.196	1.538	1.007	7.793	0.196	1.398	1.025	6.960	0.196
1973	1.282	1.052	7.727	0.158	1.584	1.008	9.977	0.158	1.425	1.030	8.780	0.158
1974	1.292	1.059	9.145	0.133	1.620	1.008	12.052	0.133	1.447	1.033	10.498	0.133
1975	1.306	1.063	10.572	0.116	1.651	1.006	14.101	0.116	1.469	1.034	12.210	0.116
1976	1.333	1.065	12.213	0.102	1.674	1.004	16.279	0.102	1.494	1.034	14.100	0.102
1977	1.416	1.069	15.550	0.085	1.729	0.996	20.362	0.085	1.564	1.032	17.794	0.085
1978	1.494	1.070	19.922	0.070	1.778	0.985	25.748	0.070	1.630	1.027	22.648	0.070
1979	1.617	1.063	26.765	0.057	1.812	0.973	32.781	0.057	1.712	1.017	29.621	0.057
1980	1.768	1.036	37.185	0.046	1.834	0.964	41.434	0.046	1.801	0.999	39.252	0.046

TABLE IX

Index of Insulation of All Gas-Heated Homes in SMUD's Service Area, Summer

Year	Equation (16)				Equation (17)				Fisher's ideal index of insulation = $[(1)\cdot(5)]^{1/2}$ (9)	Fisher's ideal index of area = $[(2)\cdot(6)]^{1/2}$ (10)	Fisher's ideal index of the number of dwellings = $[(3)\cdot(7)]^{1/2}$ (11)	Fisher's ideal index of total effect = $[(4)\cdot(8)]^{1/2}$ (12)
	Index of insulation (1)	Index of area (2)	Adjusted index of the number of dwellings (3)	Index of total effect = (1)÷[(2)·(3)] (4)	Index of insulation (5)	Index of area (6)	Adjusted index of the number of dwellings (7)	Index of total effect = (5)÷[(6)·(7)] (8)				
1960	1.000	1.000	1.000	1.000	1.000	1.000	1.000	1.000	1.000	1.000	1.000	1.000
1961	1.000	1.000	1.229	0.813	1.000	1.000	1.229	0.813	1.000	1.000	1.229	0.813
1962	1.000	1.001	1.485	0.673	1.000	1.001	1.485	0.673	1.000	1.001	1.485	0.673
1963	1.000	1.002	1.800	0.555	1.000	1.002	1.800	0.555	1.000	1.002	1.800	0.555
1964	1.000	1.002	2.162	0.461	1.000	1.002	2.163	0.461	1.000	1.002	2.163	0.461
1965	1.000	1.003	2.591	0.385	1.000	1.003	2.591	0.385	1.000	1.003	2.591	0.385
1966	1.000	1.004	3.107	0.320	1.000	1.004	3.107	0.320	1.000	1.004	3.107	0.320
1967	1.000	1.006	3.711	0.268	1.000	1.005	3.711	0.268	1.000	1.006	3.711	0.268
1968	1.000	1.007	4.079	0.244	1.000	1.006	4.080	0.244	1.000	1.007	4.080	0.244
1969	1.000	1.008	4.490	0.221	1.000	1.008	4.491	0.221	1.000	1.008	4.490	0.221
1970	1.000	1.010	4.978	0.199	1.000	1.009	4.980	0.199	1.000	1.010	4.979	0.199
1971	1.003	1.011	5.402	0.184	1.004	1.011	5.411	0.184	1.003	1.011	5.406	0.184
1972	1.007	1.013	5.891	0.169	1.010	1.012	5.912	0.169	1.008	1.012	5.902	0.169
1973	1.010	1.014	6.162	0.162	1.016	1.013	6.206	0.162	1.013	1.013	6.184	0.162
1974	1.013	1.016	6.469	0.154	1.021	1.014	6.533	0.154	1.017	1.015	6.501	0.154
1975	1.015	1.017	6.848	0.146	1.024	1.016	6.923	0.146	1.019	1.016	6.885	0.146
1976	1.017	1.019	7.220	0.138	1.027	1.017	7.307	0.138	1.022	1.018	7.264	0.138
1977	1.031	1.021	7.652	0.132	1.041	1.018	7.752	0.132	1.036	1.020	7.702	0.132
1978	1.042	1.023	8.026	0.127	1.054	1.020	8.145	0.127	1.048	1.021	8.085	0.127
1979	1.055	1.025	8.441	0.122	1.068	1.021	8.578	0.122	1.062	1.023	8.509	0.122
1980	1.064	1.026	8.810	0.118	1.079	1.022	8.971	0.118	1.072	1.024	8.890	0.118

Equation (16) can be seen to be the same as (18) by dividing all the numerators and denominators in (16) by $\Sigma \, D_{ij_0} = \Sigma \, E_{ij_0} N_{ij_0} I_{A_{ij_0}}$ and substituting from (6), (7), and (8) in the resulting expression.

Similarly, by dividing the numerators and denominators of (17) by $\Sigma \, D_{ij_t}$, (17) can be shown to be equivalent to an expression similar to (18) with s_{ij_0} replaced by s_{ij_t}.

The first term in (16) is a base-weighted index of insulation with the number of dwellings held constant at their base-year level. The second term is a current-weighted $(I_{A_t})^{-1}$, i.e., the reciprocal of the aggregate square footage of dwellings with the weights (energy required per dwelling and number of dwellings) fixed at their current year's level. The third term is an adjusted base-weighted $(I_{N_t})^{-1}$, i.e., the reciprocal of the index of the aggregate number of homes, with the weights (energy requirement and square footage) held constant at their base-year level and with the adjustment factor being the reciprocal of the microinsulation indices $I_{IN_{jt}}$.

Equation (17) provides an alternative decomposition of I_{TOT_t}. The indices are the same as in (16) except that (17) reverses the time subscripts of the weights used in (16). Where we had the reciprocal of a Laspeyres (i.e., a backward Paasche) index in (16), we now have the reciprocal of a Paasche (a backward Laspeyres) index in (17), and so on.

The insulation indices in (16) and (17) are shown in columns 1 and 5 of Table VIII for electrically heated homes and in columns 1 and 5 of Table IX for gas-heated homes. In general, when the base-year and current-weighted indices are calculated for short periods of time under either decomposition, (16) or (17), they tend to convey essentially the same picture. This has been indeed our experience with most indices of the aggregates we calculated. However, in some instances, these indices do diverge even for short periods. Moreover there are times when the base-weighted index shows a more rapid growth than the current-weighted index and the reverse tends to occur in other cases. (For a discussion of the statistical relationship between the base-weighted and current-weighted indices, see Section 2.7 in Allen, 1975.)

Even though these problems did not turn out to be troublesome in our case, one finds it desirable to have a single index of insulation, i.e., a single decomposition of the I_{TOT} in (11)—so much is accounted for by the effect of insulation, so much by the effect of area, etc., rather than

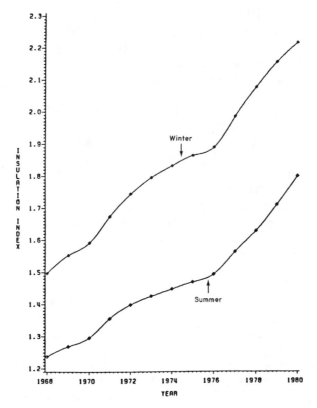

Fig. 1. Fisher's ideal insulation index of electrically heated dwellings (SMUD's service area, 1968–1980).

having two alternative splits.[13] One way of accomplishing this is to calculate Fisher's ideal index, which replaces every pair of base- and current-weighted elementary indices by the geometric average of the two indices. As we noted, for every base-weighted index in (16) the corresponding index in (17) is a current-weighted index of the same

[13] For further discussion of the problem of the uniqueness of the decomposition, see Stuval (1957). Stuval also proposed an alternative index which has the desirable properties of Fisher's ideal index (see text) but is suitable for additive analysis (that is, for an analysis explaining the difference rather than the ratio of the value in two different periods). Stuval's single index is derived from base-weighted indices of prices and quantities.

variable and vice versa. Therefore, if we replace every pair of corresponding indices in (16) and (17) by the geometric average, we have

$$I_{\text{TOT}t} = (\ell_{I_{\text{IN}_t}} p I_{I_{\text{IN}_t}})^{1/2} \times (p_{I_{A_t}} \ell_{I_{A_t}})^{-1/2} \times (Adj\ell_{I_{\text{N}_t}} Adj p_{I_{\text{N}_t}})^{-1/2}, \quad (19)$$

where ℓ and p denote Laspeyres and Paasche indices, respectively.

Column 9 of Tables VIII and IX lists Fisher's ideal indices of insulation; column 9 corresponds to the indices in columns 1 and 5 of each table. Figures 1 and 2 plot these indices. In estimation, we used only Fisher's ideal indices. As we indicated, we do not make use of the area and number-of-dwellings indices in this study. They fall off as a by-

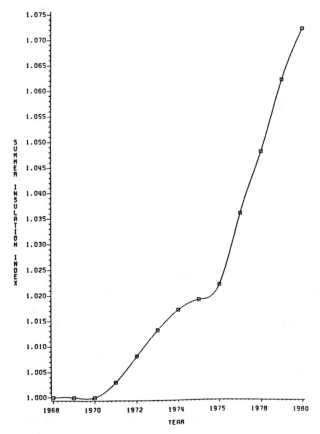

Fig. 2. Fisher's ideal index of summer insulation for gas-heated dwellings (SMUD's service area, 1968–1980).

product of the decomposition of $\Sigma\ D_{ij_0}/\Sigma\ D_{ij_t}$. But for completeness, they are listed in Tables VIII and IX.

Fisher's ideal index of insulation has certain properties not possessed by its constituent Laspeyres and Paasche indices. In particular, Fisher's ideal index meets the time-reversal test and the factor-reversal test. Briefly these tests require the following: Consider a two-factor index (e.g., price P and quantity Q). The time-reversal test requires that when the time subscripts attached to the two variables in the quantity index are interchanged, the resulting quantity index should be the reciprocal of the original expression. The same is true for the price index when the time-reversal test is applied. The factor-reversal test requires that when P and Q appearing in a price (quantity) index are interchanged, the resulting quantity (price) index will be such that when multiplied by the original price (quantity) index, it produces the value index, defined as $\Sigma_i\ P_{i_t}Q_{i_t}/\Sigma_i\ P_{i_0}Q_{i_0}$. The ability of an index to pass the factor-reversal test also makes it suitable for an additive analysis. (The fact that our index is a three-factor rather than a two-factor index poses no problem. For a demonstration that the Fisher ideal indices we calculated pass the time-reversal and factor-reversal tests, see Khazzoom, 1986, p. 73.)

IV. ESTIMATION PROBLEMS

A. Functional Form and Related Estimation Problems

1. The Box–Cox Transformation

For empirical implementation, the relationship between insulation demand and its determinants must be couched in a specific functional form. Economic theory provides little guidance as to the functional form appropriate for the relationships of our model. For convenience, we specified (1) and (2) in a linear form. Economists sometimes specify the partial adjustment model as linear in the logarithms. Either form can probably be used as an approximation. It should be obvious, however, that constraining the functional specification to take one particular form can lead to erroneous results when insufficient *a priori* information is available. Some economists (see, Spitzer, 1976) have argued for the use of the least-restrictive form.

A procedure for handling the problem of functional specification, particularly when the functional form is not suggested by theory, is the transformation of variables suggested in the early 1960s by Box and Cox (1964). Since this procedure was proposed by is authors, the procedure has been applied in econometric work in the estimation of single-equation models (e.g., Chang, 1977; Spitzer, 1976), simultaneous equation models (e.g., Spitzer, 1977), and models with truncated dependent variables (Poirier, 1978); in the choice of the form of the error distribution (Leach, 1975); and in the differentiation between the problem of serial correlation and the problem of functional specification in statistical tests (Savin and White, 1978).

The Box–Cox procedure yields a transformation which is the result of estimation rather than the subjective judgment of the investigator when insufficient *a priori* information is available. This transformation family is by no means exhaustive for finding the true model. Nonetheless, it incorporates many of the linear-in-parameter models commonly used by economists. (Applebaum also showed how the Box–Cox generalizes some of the flexible functional relationships that have been used in econometric analysis. See Applebaum (1979).

The transformation yields an estimate of the transformation parameter which enables us to discriminate between alternative models. Perhaps because of this attribute, the Box–Cox transformation has been found by some to yield models with superior forecasting performance compared to models whose functional forms were restricted *a priori*. See, for example, the results of Spitzer's study of the demand and supply of money in Spitzer (1977). See also the results of Spitzer's Monte Carlo study of the Box–Cox transformation in small samples in Spitzer (1978), particularly pp. 491–493.

Suppose that the demand for insulation is known to depend on a vector X, i.e., $I_{IN} = f(X)$, but that the specific form of this functional dependence is unknown. Box and Cox (1964) considered a class of transformations on the dependent variable given by

$$I_{IN}^{(l)} = (I_{IN}^l - 1)/l, \tag{20}$$

where l is a parameter to be estimated, and derived an approximate confidence region for l. (Subsequently, Bickel and Doksum argued that strictly speaking this confidence region is incorrect. See below.) Initially, Box and Cox proposed this transformation for the dependent variable only in order to induce a normal distribution. Economists extended the transformations to include the independent variables as well:

$$I_{\text{IN}}^{(l)} = \sum_j b_j X_j^{(l)} + u. \tag{21}$$

The functional form (21) can be used as a generalized functional specification of the relationship between the dependent variable and its determinants. The maximum likelihood (ML) method can be used to estimate the parameters of (21).

As Leech (1975) points out, we can give a meaning to the values of l, at least when l falls within the interval $[-1, 1]$, namely, that the systematic part of (21) represents some conditional measure of central tendency. When $l = 1$, (21) represents a linear relationship and the systematic part is interpreted as the arithmetic mean of the dependent variable conditional on the independent variables. When $l = -1$, the relationship is reciprocal and the systematic part would be a conditional harmonic mean. When $l = 0$, the transformed variables in (21) are not defined (0/0), but their limit as $l \to 0$ is the logarithm of the original variables. (See Kmenta (1975), p. 467. Alternatively, one may derive this result by l'Hopital's rule.) The systematic part can then be interpreted as the conditional geometric mean of the dependent variable. Similarly, when $l = 0.5$, the systematic part of (21) may be interpreted as an estimate of $E(I_{\text{IN}_t}^{0.5})$, and so on.

The linear (or loglinear) equation in (3) implies that the estimated l is constrained to be 1 (0 in the loglinear case). There is no *a priori* reason why this constraint should be imposed, and we could let our data determine the appropriate ML estimate of l.

In fact, we need not constrain the estimated l to be the same for all variables. We can gain a greater flexibility by letting the data determine the ML estimate of l that transforms each one of the variables in the model. Such a generalization enriches the family of specification alternatives. (Such specification was proposed in the early 1960s by Mukerji (1963) and Dhrymes and Kurz (1964) to generalize the production function, except that transformation was confined to raising the variable to a power l without subtracting 1 and dividing by l. This is the type of transformation proposed earlier by Tukey.) But while feasible, the generalization is costly to apply, since it requires an enormous amount of computational time. We therefore estimated our model by using the same l to transform all the variables of the model.

Some Results Pertaining to the Box–Cox Transform.[14] Several limitations of the Box–Cox transformation were brought up in the literature. In estimating the model, we allowed for those limitations that can be allowed for. Here are the major ones.

1. Schlesselman (1971) showed that the estimates of the Box–Cox transformation are not invariant to the units of measurement when a constant term is not included in the equation. (Zarembka reports that he obtained nonsensical results when he attempted to estimate a CES production function by using the Box–Cox transformation but without adding an intercept to the equation. See Zarembka, 1974, p. 100.) We added an intercept term to all equations we estimated with a Box–Cox transformation. Incidentally, Schlesselman proposed several alternative transformations that are continuous at $l = 0$ and that do not require a constant term to yield ML estimates that are scale invariant. See Schlesselman, 1971, p. 310.

2. The fundamental assumption made by Box and Cox is that the random terms in the equation of the transformed variables are independent and normally distributed (with constant variance). But for this result to hold the range of the transformed dependent variable in (21) must lie between $-\infty$ and $+\infty$. However, for some values of l, this may not be possible,[15] and normality can only be approximated. Draper and Cox addressed this question and inquired how useful the estimate of l thus obtained would be. Their results show that as long as the error term is reasonably symmetric, the ML estimate of l is robust to nonnormality and that in general the Box–Cox transformation can help "regularize" the data even when normality is not achieved (Draper and Cox, 1969).

   Amemiya and Powell (1981) addressed a related aspect of the problem posed by the truncated nature of the transformed variable. They compared the Box–Cox maximum likelihood esti-

---

[14] This discussion is technical. The reader may turn to Section IV.B without losing continuity.

[15] Since I_{IN} is positive, the transformed demand in (20) can have a normal distribution only if $l = 0$ (Hinkley, 1975, p. 101, incorrectly extends the condition to include the case when l^{-1} is an integer). In general, the $I_{IN}^{(l)}$ in (20) is bounded above by $-1/l$ when $l < 0$ and below by $-1/l$ when $l > 0$.

mates (BCMLE) with the nonlinear two-stage (NL2S) least-squares estimates and the true maximum likelihood estimates (MLE) under the assumption that the untransformed dependent variable follows a two-parameter gamma distribution. Their results were not unequivocal in regard to the choice between BCMLE and NL2S estimates. However, the NL2S estimates appeared to have an edge over the BCMLEs of their model.

It should be noted that the BCMLEs that Amemiya and Powell compared with NL2S estimates are IOLS estimates (under appropriate definition of the dependent variable). We will turn to estimation using the NL2S estimates in Section VI.E below. Anticipating results, we cannot, in general, find a finite l that optimizes the NL2S criterion because of the scale problem introduced by the Box–Cox transformation. In short, the estimator proposed by Amemiya and Powell does not exist.

3. The Box–Cox formulation assumes that the error term in the transformed equation is homoscedastic. Zarembka inquired into the question of what happens when the error term is heteroscedastic and showed that under heteroscedasticity, the ML estimate of l will not be consistent (Zarembka, 1974, p. 92).

However, Zarembka also argued that the estimated l will be biased in the direction required for the transformed dependent variable to be more nearly homoscedastic, i.e., the ML estimate of l will be biased toward stabilizing the error variance. Implicitly, however, this result requires the expected value of the log of the dependent variable to be a monotonically increasing series (see Eq. (2.2) in Zarembka, 1974, p. 92). Otherwise, there is no telling which direction the bias will go or how important it will be. Indeed, Zarembka's empirical results confirm just that. They suggest that the robustness of the estimated l to heteroscedasticity depends on the empirical problem.

Zarembka proposed a procedure for obtaining consistent estimates of l under heteroscedasticity. The procedure is based on an enormous amount of approximation.

4. Box and Cox estimated the transformed model by using maximum likelihood methods. They also proposed that in conducting the statistical inference on the estimated parameters, the transformed model should be treated as a linear model. This implies that the estimate $\hat{l}$ of the transformation parameter should be treated as if it were preassigned (ignoring the fact that it is jointly

estimated with the rest of the model's parameters and is hence correlated with the rest of the estimates). This is also what economists and statisticians have been doing for a long time, namely, first estimate a linear, loglinear, etc., version of equally plausible models. Once a functional form has been chosen, confidence intervals are calculated, given that particular functional form. But there is controversy in the literature on this point.

Using a model in which the dependent variable is the only transformed variable, Bickel and Doksum (1981) found that with a small to moderate error variance, the asymptotic marginal variances of the estimated parameters are much larger than the asymptotic conditional variances when l is known. Wong (1984) found that similar results hold when the independent variable only is transformed. Both studies conclude that in constructing intervals one should add the necessary extra terms [see (35) below] to the conditional variances of the estimated parameters. Otherwise, the stability of these estimates will be exaggerated.

Carroll and Ruppert (1982), on the other hand, found out that for a model such as (21) in which both the dependent variable and the sum of the deterministic part are transformed by the same parameter, the inflation of the variance of the estimate is not as serious as Bickel and Doksum and Wong found and that asymptotically the ratio between the marginal variance and the conditional variance is at most $\pi/2$.

Using the usual estimator of the variance, Doksum and Wong (1983) and Carroll (1985) found that the effect of using the procedure recommended by Box and Cox for certain hypothesis testing of the linear model parameters is not as severe as Bickel and Doksum (1981) and Wong (1984) reported for confidence interval estimates.

Using Box and Cox's recommended procedure of conditioning on a known l and, alternatively using Bickel and Doksum's procedure of correcting for the variance, Carrol and Ruppert (1981) and Wong (1984) also did asymptotics and Monte Carlo simulations for prediction. Their results show that the variance of the prediction is close for both procedures. In general, the ratio of the variances of the two procedures was of the order of 1–2, but typically it was close to 1.

Box and Cox (1982) rebutted Bickel and Doksum's (1981) conclusion. Box and Cox agree with Bickel and Doksum's findings that the marginal variance of the estimates is larger than corresponding conditional variance, but they question the scientific relevance of this finding. Box and Cox argued that it is not sensible scientifically to state a conclusion as a number measured on an unknown scale: if $\hat{l}$ is not given a preassigned value, the numerical value of the regression coefficient could be virtually anything. What is needed instead, they argue, is a greater understanding of the ability of particular sets of data to provide information about the class of transformation as well as a greater understanding of the circumstances under which confidence intervals estimated as if $\hat{l}$ were preassigned provide an adequate approximation. In a rejoinder, Bickel and Doksum (1983) questioned the methodological underpinnings of Box and Cox's rebuttal.

Where does that leave us? One's instinct is to take the stricter way out and estimate with the marginal asymptotic variance as Bickel and Doksum (1981) and Wong (1984) do. We did do that, and we report our results in Section V.C. Yet the real question is How meaningful are these results? Two points trouble us.

1. The size of the sample and the number of explanatory variables: Wong (1984) and Bickel and Doksum (1981) derived their results for a small model, one explanatory variable and only one variable transformed. Problems caused by the extension to four or five explanatory variables (e.g., correlation among explanatory variables, different scaling of explanatory variables, etc.), combined with problems posed by a sample of the small size with which we have been dealing, are not dealt with in these studies. Spitzer (1978) did a Monte Carlo study with $T = 30$ and $T = 60$ and applied a transformation to both dependent and independent variables. His system included two explanatory variables X_1 and X_2. The choice of X_1 and X_2 was constrained so that

$$b_1^2 \, \mathrm{Var}(X_1^{(l)}) = b_2^2 \, \mathrm{Var}(X_2^{(l)})$$

in order to equalize the contribution to the variance of $Y^{(l)}$. This is very different from any structural relationship that a model like ours could possibly have.

In short, practically no work has been done to shed light on the applicability of the results using the marginal variance of the estimates to a model similar to ours and to a sample size as small as ours.

2. The heterogeneity of the parameters for which a marginal variance is calculated: When either the dependent or independent variable or both are transformed, the regression coefficient will have a different unit of measurement (and a different meaning)

every time the transformation parameter takes on a different value. Calculating the marginal variance of a regression coefficient requires us to aggregate magnitudes that are not comparable to begin with. This problem is also related to the scale problem discussed by Box and Cox (1982).

The problem is a basic one. It will not go away even if the economist were to impose *a priori* constraints on the range of $\hat{l}$ by, say, letting $\hat{l}$ vary between -1 and $+1$ before calculating the marginal variance. Even within this narrower interval (indeed, whenever $\hat{l}$ is not given a preassigned value), the difficulty with parameter heterogeneity will arise. Neither will the problem go away if we normalized all variables (by dividing them by, say, their geometric average) to convert both the dependent and independent variables to pure numbers. The remedy is only temporary, because we eventually have to back-transform to the original units. Then the problem will reappear. Moreover, even with both the dependent and independent variables transformed into pure numbers, Box and Cox's objection to the use of the marginal variance still holds. When the transformation parameter varies over the whole real line, the numerical value of the regression coefficient could be virtually anything. A large marginal variance will follow almost by definition.

Two more observations:

1. Even if we abstract from the preceding two points, from the prediction point of view, it is not clear that much gain, if any, in forecast accuracy will result from the use of the marginal instead of the conditional variance. This is so because the regression coefficients may be expected to correlate negatively with the estimated transformation parameter. This is the essence of the prediction findings reported by Carrol and Ruppert (1981) and Wong (1984) and mentioned above.

2. The marginal variance which is larger than the conditional variance will probably be made to appear larger when the sample size is small. The conditional covariance of IOLS $\hat{\sigma}^2(X'X)^{-1}$ depends on $\hat{\sigma}^2$ (which is independent of the estimated parameters). The marginal covariance, on the other hand, depends on $\hat{\sigma}^2$ and on the six other estimated parameters in our model. (See Appendix A.) The additional six estimates introduce an added source of variability in the estimated marginal covariance matrix, which is inversely related to the sample size.

It is therefore doubtful whether with samples of the size normally available for economic time series it is possible to derive reliable estimates of the marginal variances.

In summary, in spite of whatever methodological reservations one may have, we find it more reasonable to rely on tests of significance based on a conditional rather than a marginal variance. But since this is a point of controversy, we do report results for both the conditional and marginal variances. As we will see, it is not so easy to derive reliable estimates of the marginal variances.

Our plan for the rest of this and the following two sections is as follows. We derive in the rest of this section the likelihood function and the asymptotic covariance matrix for the one-equation model; in the next section we discuss estimation problems. In Sections V.A and V.B we report the results of iterative least-squares estimates, conditional on a preassigned $\hat{l}$, and in Section V.C we report the result of estimating the marginal variances. In Section V.D we report the results of bootstrapping the one-equation model. The reporting of these estimates is structured similarly: First we report estimates with the preassigned $\hat{l}$ and then with unconditional $\hat{l}$. Section VI takes up consistent estimation using the limited information maximum likelihood.

2. The Likelihood Function and the Asymptotic Covariance

Denote the model in (3) by[16]

$$I_t^{(l)} = aI_{t-1}^{(l)} + Z_t^{(l)}b + u_t, \tag{22}$$

where $Z_t^{(l)}$ is the matrix of transformed explanatory variables other than the lagged value of the dependent variable that appear in (3). We delete the subscripts IN in this section only, for simplicity's sake. Assuming that u_t is approximately normal with 0 mean and $\sigma^2 I_T$ covariance, the likelihood of (a, b, l, σ^2) given $I_0, I_1, \ldots, I_t$, and Z_t is

$$L(a, b, l, \sigma^2 \mid I_0, I_1, \ldots, I_t, Z_t)$$

$$= (2\pi\sigma^2)^{-T/2} \exp\left\{-\frac{1}{2\sigma^2} \sum (I_t^{(l)} - aI_{t-1}^{(l)} - Z_t^{(l)}b)^2\right\} |J|, \tag{23}$$

---

[16] This is a technical discussion. The reader may turn to Section IV.B without losing continuity.

where

$$J(u_t \to I_S) = \left| \frac{\partial u_t}{\partial I_S} \right| = \left| \frac{\partial u_t}{\partial I_S^{(l)}} \times \frac{\partial I_S^{(l)}}{\partial I_S} \right|$$

$$= \begin{vmatrix} 1 & 0 & 0 & \cdots & 0 \\ -a & 1 & 0 & \cdots & 0 \\ 0 & -a & 1 & \cdots & 0 \\ \vdots & \vdots & \vdots & & \vdots \\ 0 & 0 & 0 & \cdots & 1 \end{vmatrix} \cdot \begin{vmatrix} I_1^{l-1} & 0 & 0 & \cdots & 0 \\ 0 & I_2^{l-1} & 0 & \cdots & 0 \\ \vdots & \vdots & \vdots & & \vdots \\ 0 & 0 & 0 & \cdots & I_t^{l-1} \end{vmatrix}$$

$$= \prod_{t=1}^{T} I_t^{l-1}, \tag{24}$$

which would have been the same as the Jacobian of the transformation had $I_{t-1}^{(l)}$ not appeared in (22). (If I_t is normalized such that $\prod_{t=1}^{T} I_t = 1$, then $|J| = 1$. Maximization of (23) with the I_t's now normalized is the same as minimization of the sum of the squares in the exponent of (23) given l. We can then vary the value of l over an appropriate range to find the optimal l. We will return to this subject later.

With the preceding derivations, we can treat the lagged variable just as another explanatory variable and rewrite (22) in standard matrix notation as

$$Y^{(l)} = X^{(l)}a + u, \tag{25}$$

$$u \sim N(0, \sigma^2 I), \tag{26}$$

where a is now the parameter vector. If we assume that l is known *a priori*, the *conditional* variance–covariance matrix of the estimate of a is

$$\text{Cov}(\hat{a}) = \sigma^2 (X^{(\hat{l})'} X^{(\hat{l})})^{-1}, \tag{27}$$

which converges to $\sigma^2 (X^{(l)'} X^{(l)})^{-1}$ since $\hat{l}$ is approximately consistent. (Here $\hat{l}$ is consistent only if we use the likelihood function of a truncated normal distribution and take into account the fact that $Y_t^{(l)}$ is bounded below by $-1/l$ for $l > 0$ are bounded above by $-1/l$ for $l < 0$ because $Y_t > 0$. See Bickel and Doksum, 1981.)

On the other hand, if we view $\theta = (a, l, \sigma^2)$ as unknown parameters to be estimated simultaneously by maximizing the log likelihood

$$L^* = \frac{T}{2} \log(2\pi) - T \log \sigma$$

$$- \frac{1}{2\sigma^2} (Y^{(l)} - X^{(l)}a)'(Y^{(l)} - X^{(l)}a) + (l - 1) \sum_{t=1}^{T} \log Y_t, \qquad (28)$$

then

$$\sqrt{T}(\hat{\theta} - \theta) \overset{\text{a.s.}}{\sim} N[0, (I_T/T)^{-1}], \qquad (29)$$

where I is the information matrix

$$I_T = \sigma^{-2} \begin{bmatrix} X^{(l)\prime}X^{(l)} & X^{(l)\prime}b & 0 \\ b'X^{(l)} & C & B \\ 0 & B' & 2T \end{bmatrix}, \qquad (30)$$

where b is the $T \times 1$ vector

$$b = E \frac{-\partial Y_t^{(l)}}{\partial l} + \sum_{j=1}^{p} \frac{\partial X_{tj}^{(l)}}{\partial l} a_j$$

$$= \frac{-E(V_t \ln V_t)}{l^2} + \sum_{j=1}^{p} \frac{X_{tj}^l \ln X_{tj}^l a_j}{l^2} \qquad (t = 1, ..., T), \qquad (31)$$

$$B = \frac{-2\sigma}{l} \sum_{t=1}^{T} E(\ln|V_t|), \qquad (32)$$

$$C = \sum_{t=1}^{T} \left\{ E \left(\frac{V_t \ln |V_t| - (V_t - 1)}{l^2} - \sum_{j=1}^{p} \frac{X_{tj}^l \ln X_{tj} - (X_{tj} - 1)}{l^2} a_j \right)^2 \right.$$

$$+ E \left(\frac{V_{t-1}}{l} - \sum_{j=1}^{p} \frac{(X_{tj}^l - 1)a_j}{l} \right)$$

$$\left. \times \left(\frac{lV_t(\ln|V_t|)^2 - 2l(V_t \ln|V_t| - V_t + 1)}{l^4} \right) \right\} \qquad (33)$$

with

$$V_t = Y_t^l \sim N \left(1 + l \sum_{j=1}^{p} X_{tj}^{(l)} a_j, l^2\sigma^2 \right). \qquad (34)$$

The elements of (30) are derived in Appendix A.

The asymptotic covariance of $(\hat{a})$ is

$$\text{Avar}(\hat{a}) = \sigma^2[(X^{(l)'}X^{(l)})^{-1} + FG^{-1}F'], \tag{35}$$

where

$$G = H - D'(X^{(l)'}X^{(l)})^{-1}D, \qquad F = (X^{(l)'}X^{(l)})^{-1}D,$$

$$H = \begin{bmatrix} C & B \\ B' & 2T \end{bmatrix}, \qquad D = [X^{(l)'}b \quad 0].$$

The second term on the right-hand side (35) is nonnegative definite. Hence the variances of the components of $\hat{a}$, as calculated from (35), are at least as large as the corresponding variances in (27), where $\hat{l}$ is assumed to be known *a priori*.

To calculate (31)–(33) we can either compute the approximate expectations by numerical integration or approximate these expressions by the corresponding finite sample moment estimates. We will report results using both approximations.

To facilitate computations, we divide all the variables except for the intercept by the geometric mean $\dot{Y} = (\Pi_{t=1}^{T} Y_t)^{1/T}$ to get new variables $Y_t^* = Y_t/\dot{Y}$, $X_{tj}^* = X_{tj}/\dot{Y}$. By fitting an equivalent model

$$Y^{*(l^*)} = X^{*(l^*)}a^* + \mu^*, \tag{36}$$

the log-likelihood function to be minimized becomes

$$L^* = -T/2 \log(2\pi) - T \log \sigma$$

$$- 1/2\sigma^2(Y^{*(l^*)} - X^{(*(l^*)}a^*)'(Y^{*(l^*)} - X^{*(l^*)}a^*). \tag{37}$$

Thus for a fixed l^*, we have reduced the problem to that of an ordinary least-squares problem. With this formulation, the maximum likelihood estimate for l^* is that l^* which minimizes the sum of the squares of the errors

$$S^* = (Y^{*(l^*)} - X^{*(l^*)}\hat{a}^*(l^*))'(Y^{*(l^*)} - X^{*(l^*)}\hat{a}^*(l^*)), \tag{38}$$

where $\hat{a}^*(l^*)$ is the least-squares estimate for a^* with l^* fixed. In the actual computations, we use an ordinary least-squares algorithm repeatedly over a range of values for l^* and choose that $\hat{l}^*$ which gives the highest R^2.

As it is, all the elements of a^* except for the intercept in (36) are identical to the corresponding elements of a in (25). The correspondence between the new parameters a^* and the old parameters a can be found as follows.

From (36), we have

$$\frac{(Y_t/\dot{Y})^{l*} - 1}{l^*} = a_1^* + \sum_{j=2}^{p} a_j^* \frac{(X_{tj}/\dot{Y})^{l*} - 1}{l^*} + \mu_t^* \leftrightarrows \frac{Y_t^{l*} - 1}{l^* \dot{Y}^{l*}} + \frac{1 - \dot{Y}^{l*}}{l^* \dot{Y}^{l*}}$$

$$= a_1^* + \sum_{j=2}^{p} a_j^* \left(\frac{X_{tj}^{l*} - 1}{l^* \dot{Y}^{l*}} + \frac{1 - \dot{Y}^{l*}}{l^* \times \dot{Y}^{l*}} \right) + \mu_t^* \leftrightarrows \frac{Y_t^{l*} - 1}{l^*}$$

$$= \left[(a_1^* \dot{Y}^{l*}) - \frac{1 - \dot{Y}^{l*}}{l^*} + \sum_{j=2}^{p} a_j^* \frac{1 - \dot{Y}^{l*}}{l^*} \right]$$

$$+ \sum_{j=2}^{p} a_2^* \left(\frac{X_{tj}^{l*} - 1}{l^*} \right) + \dot{Y}^{l*} \mu_t^*. \tag{39}$$

Equating the last expression of (39) with (25), we find that

$$a_1 = a_1^* \dot{Y}^{l*} + \left(\sum_{j=2}^{p} a_j^* \frac{1 - \dot{Y}^{l*}}{l^*} - \frac{1 - \dot{Y}^{l*}}{l^*} \right),$$

$$a_j = a_j^*, \quad j = 2, \ldots, p; \quad l = l^*, \quad \sigma = \dot{Y}^{l*} \sigma^*. \tag{40}$$

Thus in the asymptotic variance–covariance of $\hat{a}^*$, $\hat{l}^*$, $\hat{\sigma}^*$, the parts corresponding to the estimators $\hat{a}^*, j = 2, 3, \ldots, p$, and $\hat{l}^*$ are the same as the parts corresponding to the estimators $\hat{a}_j$ and $\hat{l}$. To perform inferences on these parameters, we need not back-transform to the original model. [If we wish to find the asymptotic variance of the intercept $\hat{a}_1$, we can invoke a well-known theorem which states that if $\theta_T \overset{\text{a.s.}}{\sim} N(\theta, I^{*-1})$, then for $X(\cdot) \in C^1$, $X(\theta) \overset{\text{a.s.}}{\sim} N[X(\theta), (\partial X(\theta)/\partial\theta)I^{*-1}(\partial X(\theta)/\partial\theta)']$. Spitzer (1982), p. 310, derives a similar relationship.] But we will not be concerned with the variance of the intercept in this study.

This procedure also simplifies the structure of the covariance (30). In particular the matrix corresponding to B in (32) reduces to the null matrix.

The information matrix in (30) becomes

$$I^* = \sigma^{-2} \begin{bmatrix} X^{*(l)'}X^{*(l)} & X^{*(l)'}b^* & 0 \\ b^{*'}X^{*(l)} & C^* & 0 \\ 0 & 0 & 2T \end{bmatrix}$$

$$= \sigma^{-2} \begin{bmatrix} P & M & 0 \\ M' & R & 0 \\ 0 & 0 & 2T \end{bmatrix} = \sigma^{-2} \begin{bmatrix} Z & 0 \\ 0 & 2T \end{bmatrix}; \tag{41}$$

then

$$(\sigma^{-2}Z)^{-1}$$

$$= \sigma^2 \begin{bmatrix} U & V \\ V' & W \end{bmatrix}$$

$$= \sigma^2 \begin{bmatrix} P^{-1} + P^{-1}M(R - M'P^{-1}M)^{-1}M'P^{-1} & -P^{-1}M(R - M'P^{-1}M)^{-1} \\ -(R - M'P^{-1}M)^{-1}M'P^{-1} & (R - M'P^{-1}M)^{-1} \end{bmatrix}.$$

$$\tag{42}$$

Hence, (35) simplifies to

$$\text{Avar}(\hat{a}^*) = \sigma^2(P^{-1} + P^{-1}MWM'P^{-1})$$

$$= \sigma^2[P^{-1} + (P^{-1}MW)W^{-1}(P^{-1}MW)']$$

$$= \sigma^2(P^{-1} + VW^{-1}V'). \tag{43}$$

Here $\sigma^2 V$ is the covariance vector of $\hat{a}^*$ and $\hat{l}$ and $\sigma^2 W$ is the variance of $\hat{l}$. Hence the variance calculated from OLS, when $\hat{l}$ is treated as if it were known *a priori*, underestimates the $\text{Avar}(\hat{a}_j^*)$ by an amount equal to [covariance($\hat{a}^*$, $\hat{l}$)]^2/variance($\hat{l}$). This result restates the basis of Bickel and Doksum's objection to Box and Cox's procedure for calculating the variance of the estimated parameters.

B. The Simultaneity Problem

1. Energy Price and Insulation Demand

The existence of simultaneity, at least in principle, between electricity price (or gas price) and insulation demand is traceable to the simultaneity between energy demand and energy price. If we grant for a moment the existence of simultaneity between energy demand and energy price, it follows that energy price and insulation demand must be jointly determined. This is so because energy price depends on energy demand, which in turn depends on, among other things, the insulation level. (See Khazzoom, 1986.) Hence, the price of energy which appears as a determinant of the demand for insulation in (3) is in turn determined by the demand for insulation. Since the chain of relationship hinges on the existence of a simultaneity between energy demand and energy price, we devote this section to a brief discussion of this subject.

Several authors addressed the problem of demand–price simultaneity in the estimation of electricity demand. The general proposition is that explanatory variables whose construction utilizes the observed consumption level of electricity (e.g., electricity price) are, in general, correlated with the error term. Hence their use in demand estimation creates a simultaneous-equation problem. Ordinary least-squares estimates, under the circumstances, will be inconsistent. Note that the simultaneity issue arises not only when the demand equations are estimated with average price, but with *ex post* marginal price as well.

To establish an empirical verification of the hypothesis of endogeneity, Dubin (1981) carried out a specification test of exogeneity for the average price as well as for the marginal price. (Dubin used disaggregated data for household derived from a 1975 survey of 1502 households in order to avoid the bias of aggregation and approximation of the rate structure.) His test results rejected the exogeneity of these variables. (See Dubin, 1981, pp. 12–17.) A comparison of OLS with instrumental variable estimates also showed that the direction of the bias conforms to *a priori* expectations and that moreover the bias was on the order of 67%.

The remedy for the simultaneous-equation problem is, of course, to use a simultaneous-equation estimation procedure. (This is the approach used by Halvorsen (1978). In addition to the demand equation, Halvorsen specifies a price equation.) However, the usual procedures (such as two-stage least-squares, etc.) may not be appropriate when microdata on individual customers are used, because the utility rate structure typically features price–quantity breaks rather than a smooth price–quantity relationship. Instrumental variable methods have been proposed in the literature that take into account the endogeneity of a discrete variables. (See, for example, Rosen, 1976; Burtless and Hausman, 1978; and Wales and Woodland, 1979). Wales and Woodland (1979) also allowed for the possibility that the individual customer may be on a segment of the piecewise linear constraint other than the observed one because of random disturbances.) Hausman *et al.* (1979) and McFadden *et al.* (1977) used a related instrumental variable procedure to derive consistent estimates for a model using microdemand data for electricity.

When the data represent aggregates over a large number of customers, we may expect the aggregation to produce a relatively smooth price–quantity relationship. In that case, any of the simultaneous-equation procedures can be used.

Note, however, when the tail-end block is used as the price variable, we do not need to use a simultaneous-equation procedure. Ordinary least-squares estimate will produce consistent estimates, since the tail-end price is exogenous, provided that the measurement errors are negligible.[17] The greater the proportion of customers who fall into, or close to, the tail-end block of the rate schedule, the higher the expected correlation between the marginal price and the tail-end price and the less consequential will be the measurement errors introduced by using the tail-end block instead of the marginal price.

With that in mind, we examined the distribution of SMUD's customers on the steps of the rate structure. As it turned out, during the sample period, most of these customers fell at the tail-end block.[18] Hence we use the tail-end block of the electricity and gas price in our estimation of the residential demand for insulation.

Finally, let us address the role of the fixed cost component in approximating the price of electricity (and gas). In an article surveying the literature on the estimation of the demand for electricity, Taylor (1975) argued that both the marginal price of electricity and either the average price for blocks other than the final one (intramarginal price) or the total payment for blocks other than the final one should be included in the specification of the demand equation. The inclusion of payments for blocks other than the last one is intended to capture the effect of the fixed cost of electricity, which represents a negative income effect.

In a subsequent article Nordin (1976) argued that the use of the intramarginal price or the total payment for blocks other than the final one, as suggested by Taylor, may lead to predicting the same quantities in

[17] The problem with these measurement errors is not quite the same as the classic error-in-the-variable problem, since the expected value of the measurement errors will be negative, not zero.

[18] As it is, all electrically heated homes fell far out at the tail-end block of the summer price structure in every year covered by the sample period (1969–1980). During the winter, the average electrically heated customer's consumption fell generally in the tail-end block. When that did not happen, however, the difference was small, 4–5% away from the tail-end block. In fact, even this difference may not have existed in reality. This is so because our calculation for the typical customer did not differentiate between summer and winter (which means that the average winter consumption of an electricity customer was underestimated). Any small allowance we make for this underestimate will move the typical customer to the tail-end block. For the summer, however, the picture would probably not change because the average household was far out in the tail-end block. For gas-heated homes, the average customer fell all the time at the tail-end block, except in the winter of 1969, when he was within 3% of the tail-end block.

two times or places when in fact different quantities ought to be predicted. Nordin argued that a more appropriate variable than either of the two alternatives proposed by Taylor is the rate structure premium. Rate structure premium is defined as the excess of actual cost of electricity consumed over the calculated cost when the electricity consumed is priced at the marginal block rate.

Since the publication of Nordin's procedure, several authors, including Taylor *et al.* (1977), have used Nordin's rate structure premium instead of Taylor's intramarginal price or the total payments for blocks other than the final one in estimating the residential demand for electricity. The way in which the rate structure premium has been included in the various models is varied. Houthaker, for example, subtracted the premium from income. The idea is that since these expenditures represent a negative income effect, their parameter should be constrained to be the same as the income's. Taylor *et al.* (1977), on the other hand, estimated the parameter for the intramarginal expenditures separately; i.e., the fixed component appeared as a variable separate from income. As it turned out, the coefficient of the intramarginal expenditures in Taylor *et al.*'s study was negative, but its magnitude was only one-eighth of the income coefficient. Taylor *et al.* observed that it may very well be that people simply do not perceive their income, as reflected in their paychecks, in the same way that they perceive the pure income effect arising from price changes. (In a more recent expanded study, Taylor reports that for some models the fixed charge for electricity has the wrong signs. See Taylor *et al.*, 1982, Table 7-4, pp. 7–8.)

In a separate study, Berndt (1978) argued that the quantitative significance of Taylor's argument for including not only the marginal price, but also the average intramarginal price or the total payment for blocks other than the final one as well is negligible. Berndt argued that the least-squares estimates, with the average intramarginal price excluded from the equation, will typically differ very little from regression estimates with the intramarginal price included. To demonstrate that point, Berndt re-estimated equations developed by Houthaker in 1951 with and without the intramarginal price in the equation (in addition to the marginal price). His results were close in both cases. In light of these results, Berndt concluded that the rather costly construction of a data set for prices based on actual rate schedule is unwarranted.

Without taking issue with the merit of Berndt's empirical findings, his argument against the theoretical merit of the inclusion of the intra-marginal price is based on a wrong expression he used for the elasticity of the demand with respect to the expenditures on electricity. As it is, the theoretical bias, small as it may be, is much larger than the bias Berndt calculated. For a discussion of this point, see Khazzoom (1986).

Dubin (1981) addressed the same question raised by Berndt. As noted earlier, Dubin used disaggregated data for households derived from a 1975 survey of 1502 households. For the sample as a whole, the mean value of the rate structure premium was $3.12, compared to a mean value of monthly income of $1321.00, approximately 0.25%. In SMUD's service area, the ratio ranged from 0.5 to 0.7% in the late 1960s to the late 1970s (Sacramento Municipal Utility District, 1978, pp. E7-6, E7-9). Because of this generally small value (of the rate structure premium compared to personal income), the correction of income for the rate structure premium cannot, in general, be distinguished from other measurement errors in the definition of income. Dubin's estimation results with and without the rate structure premium subtracted from income yielded identical income (and price) parameters to the third decimal. (See Tables 2 and 4 in Dubin, 1981, pp. 16–19. See also the discussion in Judge *et al.*, 1980, Chapter 13, and Dhrymes, 1970, Chapter 5.) Similar findings were reported by Hausman *et al.* for the empirical insignificance of the rate structure premium (Hausman *et al.*, 1979, pp. 280–281).

In light of the empirical results reported by Taylor, Berndt, Dubin, and Hausman *et al.* and in light of the small share of the rate structure premium in the household's income, we do not include this variable in the electricity (or gas) price concept used in this model. In summary, for energy price, we use only the tail-end block rate deflated by the CPI.

2. Energy Usage and Insulation Demand

The simultaneity problem in the area of energy usage and insulation demand is evident. Energy usage for space heating (or air conditioning), which is a determinant of insulation demand, is in turn determined by the (resulting) insulation level. Hence, energy usage is correlated with the random variable of the model, and the OLS estimates are inconsistent.

In our model we used total energy demand instead of just energy demand for space heating or air conditioning. This does not change the nature of the problem. By using the total winter electricity use $E^W_{HH_t}$, where

$$E^W_{HH_t} = \zeta E_{SH_t} + \omega_t, \tag{44}$$

in place of (as a proxy for) E_{SH_t}, electricity use for space heating, we have introduced measurement errors which result in a nonzero correlation between $E^W_{HH_t}$ and the random term of the model. In either case, whether we use E_{SH_t} (if it is available) or $E^W_{HH_t}$, our estimates will be inconsistent. In either case, we need to use consistent methods of estimation. The main difference is that the probability limit (under the usual assumptions of asymptotic convergence of the relevant matrices) will understate the impact of space heating on insulation by a factor of ζ^{-1} (which for electrically heated homes probably ranges between 2.0 and 2.5).

Similar remarks apply to the effect of using total electricity use in the summer in place of electricity use for air conditioning or total gas use in the winter instead of gas use for space heating.

C. Operational Definitions of the Variables

The next two sections report estimation results for the period 1969–1980. The data used in estimating the insulation demand consist of observations on the residential insulation indices for

1. electrically heated dwellings during the winters and summers of 1969–1980 (these dwellings are classified in SMUD's records under rate 14) and
2. gas-heated homes during the summers of 1969–1980 (these dwellings are classified in SMUD's records under rate 17).

The following list of notation links the symbols to the data used in estimation.

I_{IN} Index of insulation level for residential dwellings in SMUD's service area, 1969, equals 1.0 (Source: Tables VIII and IX of this volume)

RTEBE   Tail-end block of electricity price deflated by CPI (Source: Tables B.I and B.II in Appendix B)

RTEBG   Tail-end block of gas price in dollars per therm, deflated by CPI (Source: Tables B.I and B.II in Appendix B)

RYPC   Income per household in SMUD's service area, deflated by CPI[18a] (Source: Tables B.II in Appendix B)

E_{HH}^S Summer electricity usage in kilowatt hours per household (Source: Tables II and III)

E_{HH}^W Winter electricity usage in kilowatt hours per household (Source: Tables II and III)

G_{HH} Annual gas usage in therms per household in Sacramento County (Source: Table B.III in Appendix B)

R_{AAA} Moody's AAA bond rate deflated by CPI (Source: Table B.II, Appendix B)

V. IOLS AND BOOTSTRAP ESTIMATION RESULTS

We report in this section the results of iterated OLS estimates for electrically heated and gas-heated homes. All of these estimates were derived by using the SAS program available on U.C. Berkeley's IBM.

A. IOLS Results for Electrically Heated Dwellings

Tables X and XI show the results of estimating for electrically heated homes. Recall that all variables except the intercept have been normalized by dividing by the geometric mean of the insulation index. Hence, except for the intercept, the estimates are identical to the estimates derived with the unnormalized data.

Each table shows one equation per line. Each column contains the estimated parameters of the variables that appear at the head of that column. The signs of the coefficient are given as though we have written the equation with the explanatory variable appearing on the right-hand side of the equality symbol. The ratio of the estimate to its standard error is shown in parentheses underneath the estimated coefficient. (We refer to these ratios as t ratios. In a simultaneous-equation model, they do not have a t distribution but are approximately

---

[18a] There are two reasons why we used real income per household rather than real income per family. One is that household corresponds more closely to the unit for which an electricity bill is issued. A family that rents one or two rooms to a tenant is recorded in the census survey as a household. Since, in general, such families have only one meter, the relevant variable is the income of all occupants of the dwelling rather than the income of the family. The second is that income per family excludes the income of households that are single.

TABLE X

OLS Estimates of the Residential Demand for Insulation for All Electrically Heated Homes in SMUD's Service Area, 1969–1980

Equation	Coeff. and t ratio	Intercept (1)	Lagged insulation (2)	Real electric price (3)		Real income per HH (4)
				(A) 1969–1980	(B) 1974–1980	
1	Coeff.	0.08701	0.98184	10.751	−3.757	−0.0004396
	t	(0.50)	(10.42)	(1.84)	(−1.20)	(−0.33)
2	Coeff.	−0.42022	0.92372	0.19813	0.01252	−0.006866
	t	(−0.44)	(11.78)	(0.43)	(2.18)	(−1.09)
3	Coeff.	−0.34362	0.90377	0.04520	0.007468	−0.01797
	t	(−0.91)	(11.90)	(0.39)	(2.29)	(−1.26)
4	Coeff.	−0.33189	0.89064	0.009683	0.003803	−0.04422
	t	(−2.08)	(12.19)	(0.33)	(2.32)	(−1.37)
5	Coeff.	−0.72047	0.88412	0.001622	0.001661	−0.10415
	t	(−1.90)	(12.61)	(0.22)	(2.34)	(−1.44)
6*	Coeff.	−2.9308	0.88229	0.0001041	0.0006323	−0.23851
	t	(−1.59)	(13.14)	(0.06)	(2.43)	(−1.46)
7	Coeff.	−14.247	0.88210	$−5.70 \times 10^{-5}$	0.0002134	−0.53828
	t	(−1.49)	(13.80)	(−0.12)	(2.58)	(−1.44)
8	Coeff.	−74.163	0.88132	$−3.25 \times 10^{-5}$	6.52×10^{-5}	−1.2112
	t	(−1.42)	(14.53)	(−0.30)	(2.76)	(−1.41)
9	Coeff.	−2298.9	0.87676	$−3.63 \times 10^{-6}$	5.01×10^{-6}	−6.2076
	t	(−1.31)	(15.97)	(−0.58)	(3.00)	(−1.34)

normal.) The standard errors used in calculating these ratios are conditional standard errors. Estimates with marginal standard errors are reported in Section V.C for model 6* in Table XI. We did not do comparable calculations with the marginal standard errors for gas-heated homes partly because of the small size of the sample for gas-heated homes.

The parameter l used in transforming the variables of each equation is reported in column 9 of each table. In estimation, we changed l in steps of 0.25.

Two columns in each table call attention to any wrong sign in the estimated parameters of the equation that has the highest R^2 (the starred equation), as well as to any coefficient in that equation that has

TABLE X (*Continued*)

Electricity use per HH (5)	Real interest rate (6)	Column with incorrect sign (7)	Correct sign but $t \leq 1$ (8)	l (9)	R^2 ($\bar{R}^2$)	SSE
7.34 × 10⁻⁶ (2.52)	−2.5583 (−2.42)			1	0.9964 (0.9951)	0.0021564
0.0004205 (2.22)	−0.52477 (−2.99)			0.5	0.9971 (0.9960)	0.0017031
0.003648 (2.23)	−0.23194 (−3.14)			0.25	0.9972 (0.9963)	0.0016019
0.02994 (2.13)	−0.10120 (−3.26)			0	0.9973 (0.9964)	0.0015310
0.22880 (2.94)	−0.04359 (−3.33)			−0.25	0.9974 (0.9965)	0.0014914
1.6313 (1.72)	−0.01858 (−3.35)	—	(3A)	−0.5	0.9965 (0.9974)	0.0014800
11.146 (1.55)	−0.007857 (−3.34)			−0.75	0.9974 (0.9965)	0.0014950
75.297 (1.44)	−0.00330 (−3.29)			−1	0.9974 (0.9964)	0.0015282
3454.6 (1.31)	0.0005779 (3.15)			1.5	0.9972 (0.9962)	0.0016578

the right sign but whose t ratio is below 1. An equation with no entry in either of these two columns conforms to the *a priori* notions about sign and has no insignificant coefficient at the indicated level.

Of the last two columns, one reports the coefficient of determination R^2 and the adjusted coefficient of determination $\bar{R}^2$; the other shows the sum of the squares of the errors (SSE). This last column is helpful in discerning among models when the changes in R^2 were too small to be reflected in the decimals printed by the program.

Table X shows the results of estimating the model in Eq. (3), which is the most restrictive model. Equation 6* with $l = -0.5$ maximizes the R^2 for the grid over which we searched for l. All signs of the estimated parameters conform to *a priori* expectations. The distinguishing fea-

TABLE XI

OLS Estimates of the Residential Demand for Insulation for All Electrically Heated Homes in SMUD's Service Area with Electricity Price from 1974 On, 1969–1980

Equation	Coeff. and t ratio	Intercept (1)	Lagged insulation (2)	Real electricity price, 1974–1980 (3)	Income per HH (4)	Use per HH (5)	Interest rate (6)	Column with incorrect sign (7)	Correct sign but $t \leq 1$ (8)	l (9)	R^2 ($\bar{R}^2$)	SSE
1	Coeff.	0.30481	0.87774	−4.8822	−0.001526	5.87×10^{-6}	−3.0106			1	0.9957 (0.9945)	0.0025880
	t	(2.25)	(10.94)	(−1.49)	(−1.21)	(1.97)	(−2.74)					
2	Coeff.	−0.81307	0.89898	0.01401	−0.008850	0.0003804	−0.55509			0.5	0.9970 (0.9962)	0.0017219
	t	(−2.96)	(17.13)	(3.13)	(−2.08)	(2.36)	(−3.53)					
3	Coeff.	−0.47356	0.88352	0.008216	−0.02192	0.003301	−0.24464			0.25	0.9972 (0.9964)	0.0016165
	t	(−2.71)	(16.22)	(3.17)	(−2.20)	(2.46)	(−3.78)					
4	Coeff.	−0.35172	0.87565	0.004117	−0.05143	0.02714	−0.10602			0	0.9973 (0.9966)	0.0015410
	t	(−2.43)	(15.60)	(3.16)	(−2.21)	(2.47)	(−3.96)					
5	Coeff.	−0.66362	0.87564	0.001751	−0.11434	0.21183	−0.04499			−0.25	0.9974 (0.9967)	0.0014957
	t	(−2.46)	(15.35)	(3.12)	(−2.11)	(2.43)	(−4.04)					
6*	Coeff.	−2.8531	0.88039	0.0006408	−0.24413	1.5943	−0.01874	—	—	−0.5	0.9974 (0.9967)	0.0014803
	t	(−2.40)	(15.60)	(3.11)	(−1.93)	(2.39)	(−4.02)					
7	Coeff.	−15.132	0.88607	0.0002075	−0.51060	11.791	−0.007703			−0.75	0.9974 (0.9967)	0.0014944
	t	(−2.41)	(16.41)	(3.13)	(−1.74)	(2.41)	(−3.94)					
8	Coeff.	−85.512	0.89070	6.14×10^{-5}	−1.0620	86.518	−0.003141			−1	0.9973 (0.9966)	0.0015360
	t	(−2.47)	(17.68)	(3.16)	(−1.57)	(2.46)	(−3.83)					
9	Coeff.	−3044.6	0.89596	4.55×10^{-6}	−4.6215	4571.4	−0.0005149			−1.5	0.9971 (0.9963)	0.0016905
	t	(−2.59)	(20.86)	(3.16)	(−1.26)	(2.59)	(−3.55)					

ture of this model is the difference between the estimated parameters for the two price variables. The model captures a strong price effect, beginning around the oil crisis period, but it shows a weak price effect in the 1969–1973 period. This is indicated both by the magnitude and by the stability of the parameter attached to the 1974–1980 price vector as opposed to the 1969–1980 price vector. We do not conclude from this result, however, that prior to 1974 electricity price played no role in determining the demand for insulation. The history of insulation of electrically heated homes at SMUD indicates that many of the insulation activities in the pre-1974 period were undertaken because SMUD's customers recognized that electric heat is expensive. But the relationship is difficult to capture *statistically* for the period prior to 1974 because real electricity price did not exhibit much variability in the 1969–1973 period.

Before we examined the effect of dropping the 1969–1980 price vector from the equation and leaving in only the 1974–1980 price vector, we attempted to estimate the quantitative impact of the utility's insulation program, which began in the second half of 1977 (The details are shown in Khazzoom, 1984, p. 88.) The estimated l that maximized the R^2 is -0.25. For this l the 1978–1980 price vector turned out to have a negative sign and an insignificant effect. But as before, the 1974–1980 price had a significant positive effect on insulation demand.

Table XI shows the effect of retaining only the 1974–1980 price vector. The estimated l that maximizes the R^2 is -0.5, the same as for Table X. The removal of the 1969–1980 price from the model hardly affects the estimates of Table X. Almost all parameters are identical to the parameters in equation 6* of Table X (they all fall within a fraction of a standard error), except that every parameter now is more stable than the corresponding parameter in Table X.

Because of the presence of another endogenous variable in the equation, E_{HH_t}, strictly we cannot calculate demand elasticities by using our present estimates. But to get a rough idea of the magnitudes involved, we have made some calculations from 6* in Table XI. These are listed in Table XII and plotted in Figs. 3 and 4 for each year in the sample period. For a Box–Cox-transformed model such as $I_{IN_t}^{(l)} = a_0 + a_1 p_t^{(l)} + a_2 I_{IN_{t-1}}^{(l)}$, the short-run elasticity with respect to price is given by $a_1(P_t/D_t)^l$, and the long-run elasticity is given by $(a_1/1 - a_2)(P_t/D_t)^l$. The elasticities shown in Table XII are the geometric averages of the elasticities plotted in Figs. 3 and 4.

Table XII shows that the short-run price elasticity for the winter is

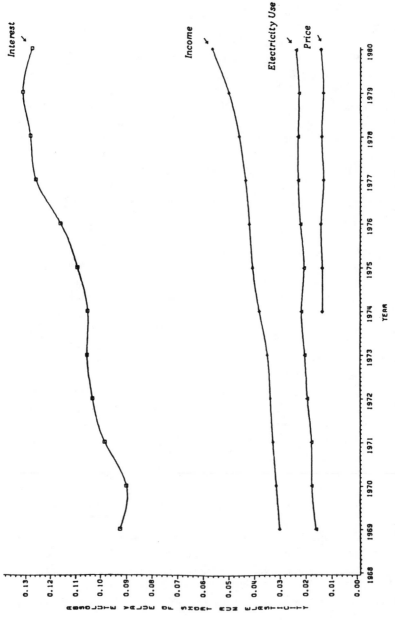

Fig. 3. Short-run winter elasticity of the residential demand for insulation of electrically heated homes (SMUD's service area, 1968–1980).

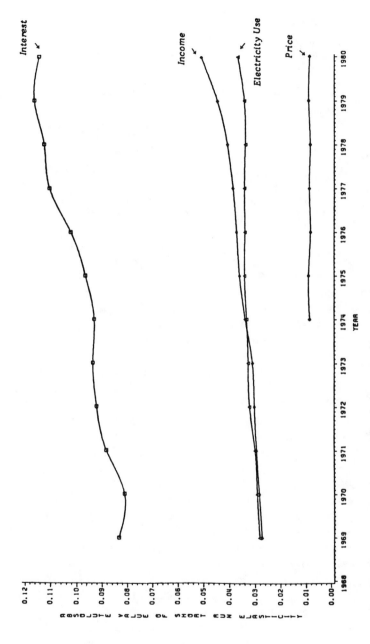

Fig. 4. Short-run summer elasticity of the residential demand for insulation of electrically heated homes (SMUD's service area, 1968–1980).

239

TABLE XII

Average Elasticities for Residential Insulation Demand Calculated from Equation 6*, Table XI

Elasticities	Winter	Summer
Short-run price elasticity	0.013	0.008
Long-run price elasticity	0.110	0.070
Short-run income elasticity	−0.038	−0.034
Long-run income elasticity	−0.325	−0.290
Short-run electricity-use elasticity	0.021	0.033
Long-run electricity-use elasticity	0.176	0.280
Short-run interest elasticity	−0.110	−0.098
Long-run interest elasticity	−0.920	−0.822

0.013 and for the summer it is 0.008, which is about two-thirds of the winter price elasticity. This is probably as it should be, since for electrically heated homes the impact of a price change will weigh more heavily when it occurs in the winter than when it occurs in the summer because of the heavier use of electricity in the winter than in the summer. Surprisingly, use-per-household elasticity shows opposite results—the elasticity is higher in the summer than in the winter. Offhand, it seems difficult to explain this result other than by the fact that it may reflect the simultaneous-equation problem, which probably affects electricity use more than the other variables that appear in our equation.

As expected, the table shows that the demand for insulation is an inferior good, with little difference between the summer and the winter income elasticity. (This is a loose statement. Strictly we should have calculated the asymptotic variance for the elasticities. It is not too difficult to do that (see, e.g., Khazzoom, 1971,), but we did not do the calculation because of the tentative nature of our results.) It is also interesting that the summer elasticities with respect to electricity usage and with respect to income are almost identical. They are four times as large as the corresponding price elasticities. The winter elasticity with respect to electricity usage and income are twice and three times as large, respectively, as the corresponding price elasticities.

But the most striking feature of the results is the interest rate elasticity of insulation demand. Of all elasticity estimates, the elasticity with respect to the interest rate is the largest—3 times as large as the income elasticity and 9 to 12 times as large as the price elasticity. (Note also the *t* ratio of the interest rate coefficient is also the highest.) This is also reflected in Figs. 3 and 4. These charts also show that the interest rate elasticity has been rising throughout the period. These results lend support to the increased emphasis that regulatory agencies and utilities alike have placed on using subsidized interest rates as a means of stimulating insulation demand. The charts also show that, like the interest elasticity, the income elasticity of insulation demand has been subject to an upward trend during the sample period. If this result has any generality, it would cast doubt on the efficacy, as well as the wisdom, of the rebates proposed by some utilities as a way of stimulating conservation demand.

We also note that the price elasticities remained almost constant, while the electricity-use elasticities appear to have been subject to only a gentle upward trend. It is also of interest to note that the price elasticity is the smallest of all elasticities we calculated. [This probably confirms the fears of many that an exclusive reliance on the price of energy will not do "enough" (whatever that means) to stimulate energy conservation via insulation.]

Two observations on the effect of income on insulation demand:

1. Some readers may question the validity of a negative income coefficient. For two reasons, they may argue the income coefficient should be positive: (1) studies of the household demand for electric appliances have uncovered an inverse relationship between income and the discount rate (see Hausmann *et al.*, 1979; and Dubin and McFadden, 1980); (2) since the return to the durables is not taxed, higher-income households earn a higher return on cost reduction than do lower-income households.

   We take no issue with (2). In fact, this factor may even more than offset the effect of the decline in the marginal utility of income. Neither do we question the validity of the result in (1). But the result in (1) was derived from the demand for electric appliances. The good we have here is not a neat electric appliance that one can pleasantly purchase in a store. Many view home insulation as a nuisance. The requirements of the process of home insulation are exacting to most households. It is this aspect that makes insulation an inferior good.

   But we went one step further. To explore the possibility that the sign of the income coefficient may in any way be related to the fact that we estimated one income coefficient for the whole period, we re-estimated with two income coefficients, one for the pre- and one for the postembargo period. The results of re-estimating the equations listed in Table XI with separate income coefficients for the 1969–1973 and 1974–1980 subperiods yielded no positive coefficient for any equation in either subperiod. Moreover, in all cases, the t ratio of the income (as well as the price) coefficient deteriorated to insignificance, perhaps because of multicollinearity. To save space, we did not tabulate these results here. We note also in passing that we obtained similar results when we re-estimated all the equations reported in Table XXIV below by using limited information maximum likelihood (LIML) methods. The only difference from the results we just reported was that the income coefficients for both subperiods 1969–1973 and 1974–1980 turned out to be almost identical. Moreover, the income coefficients were more stable with (LIML) than with IOLS, and the conditional t ratio of the coefficients to the asymptotic standard error for the 1974–1980 period never fell below 2 in all the equations we estimated.

2. In our model 6* in Table XI, we found out that the summer income elasticity and the summer electricity-usage elasticities are of about the same magnitude. Whether the net-income elasticity (that is, the sum of the direct and indirect income elasticities) is positive or negative depends on the magnitude of the short-run elasticity of the household demand for electricity with respect to income. Since the latter is generally smaller than 1, the net short-run income elasticity of insulation demand in the summer may be reasonably expected to be negative. This can be easily seen in the following simple double logarithmic model, in which for simplicity we also abstracted from the lagged values of the dependent variables.

Suppose that the equations for the demand for electricity and the demand for insulation are:

$$\ln E_t = a_0 + a_1 \ln I + a_2 \ln YPC, \tag{45}$$

$$\ln I_t = b_0 + b_1 \ln E - b_2 \ln YPC. \tag{46}$$

Substitute from (45) into (46):

$$\ln I_t = \frac{b_0 + b_1 a_0}{1 - b_1 a_1} + \frac{b_1 a_2 - b_2}{1 - b_1 a_1} \ln YPC. \tag{47}$$

Since $0 < b_1$, a_1, $a_2 < 1$ and since b_1 and b_2 are of about the same magnitude, it follows that the second term on the right-hand side of (47) is negative. A similar result holds for the winter as well as for long-run summer and winter. (To establish the long-run result, we need simply to solve a system of two simultaneous difference equations for its equilibrium value. See Goldberg, 1961, pp. 207 ff.)

The average long-run elasticities are also shown in Table XII. They are a little over eight times as large as the short-run elasticities, reflecting the fact that only a small fraction (about 12%) of the tension between the target level and the past level of insulation gets made up for. As it is, the median lag (that is, the number of years it takes for 50% of the total adjustment to be completed) is approximately 5 years. [To calculate the median lag, we write the probability-generating function (pgf) implicit in (1) as $S(B) = s \sum_{i=0}^{\infty} (1 - s)^i B^i$, where B is the backward shift operator. The standardized interim multiplier (which is the interim multiplier divided by the total multiplier) is $1 - (1 - s)^{t+1}$. To derive the median lag, we solve for t, which equates the standardized interim multiplier to 0.5, that is, $1 - (1 - s)^{t+1} = 0.5$. In our case, $s = 0.12$, which implies that t is 4.45 years. We added 0.5 years to ensure the equality of the median and the mean lag for a symmetric distribution.] The mean lag is 7.3 years, which is 1.5 times as large as the median lag, because the distribution is skewed to the right. (The mean lag is derived by differentiating the pgf with respect to the backward shift operator B and evaluating at $B = 1$. Since our results in 6* of Table XI yield $s = 0.12$, the pgf of our geometric lag structure is $0.12/1 - 0.88B$, which on differentiating and setting $B = 1$ yields an average lag of $0.88/0.12 = 7.3$ years.) The mean lag is not as useful as the median lag, because it does not tell us what percentage of the lag will be completed in 7.3 years, although we can calculate that separately if we want to.

We turn now to the estimates for gas-heated homes.

B. IOLS Results for Gas-Heated Dwellings

Table XIII shows the results of our iterated OLS estimates for gas-heated dwellings. The results pertain to the summer only. Here we used gas price and annual gas consumption per household (instead of electricity price and summer electricity consumption per household) on the theory that for gas-heated homes, the decision to insulate is driven more by gas price and gas usage than by electricity price and

TABLE XIII

OLS Estimates of Summer Demand for Insulation for Gas-Heated Homes in SMUD's Service Area, 1969–1980

Equation	Coeff. and t ratio	Intercept (1)	Lagged insulation (2)	Real gas price (3) (A) 1969–1980	Real gas price (3) (B) 1974–1980	Real income per HH (4)
1	Coeff.	0.55392	0.47462	28.901	−10.914	−0.0001576
	t	(2.54)	(2.16)	(3.51)	(−3.11)	(−1.05)
2	Coeff.	1.0268	0.49434	0.60789	0.004497	−0.001789
	t	(2.36)	(2.18)	(2.93)	(3.36)	(−1.24)
3	Coeff.	0.29289	0.50080	0.11251	0.002854	−0.005411
	t	(1.86)	(2.21)	(2.91)	(3.53)	(−1.28)
4	Coeff.	0.10455	0.50460	0.02027	0.001429	−0.01637
	t	(1.17)	(2.22)	(2.86)	(3.70)	(−1.34)
5	Coeff.	0.008353	0.50732	0.003516	0.0005426	−0.04942
	t	(0.06)	(2.24)	(2.75)	(3.86)	(−1.41)
6	Coeff.	−0.34551	0.51217	0.0005919	0.0001616	−0.14680
	t	(−0.72)	(2.29)	(2.58)	(4.00)	(−1.49)
7	Coeff.	−13.560	0.53256	1.67×10^{-5}	9.27×10^{-6}	−1.1917
	t	(−1.43)	(2.48)	(2.21)	(4.07)	(−1.64)
8	Coeff.	−427.95	0.54444	5.15×10^{-7}	4.25×10^{-7}	−9.2825
	t	(−1.88)	(2.69)	(2.09)	(3.85)	(−1.85)
9*	Coeff.	−2417.8	0.54223	9.288×10^{-8}	8.866×10^{-8}	−26.275
	t	(−2.13)	(2.79)	(2.167)	(3.80)	(−2.01)
10	Coeff.	9.9035	0.84349	2.649×10^{-9}	8.525×10^{-9}	−19.826
	t	(0.60)	(4.70)	(.47)	(2.56)	(−0.60)

a Model not of full rank.

summer electricity usage. In estimating for gas-heated homes, we defined the post-oil-crisis era to begin in 1975 rather than 1974 to allow for the delayed response of gas use to the oil crisis.[19]

[19] In the case of electricity, the response was almost immediate. Many did not use Christmas lights in December 1973. Gas price was regulated tightly at the time. Unlike electricity, whose cost of production may increase with an increase in oil price, many felt that gas price would be affected much less than electricity by the oil price increase. In fact, the FPC did not sanction the hike that tripled the price of new gas until July 1976.

TABLE XIII (*Continued*)

Gas use per HH (5)	Real interest rate (6)	Column with incorrect sign (7)	Correct sign but $t \leq 1$ (8)	l (9)	R^2 ($\bar{R}^2$)	SSE
4.08×10^{-6} (−0.52)	−0.53514 (−4.04)			1	0.9984 (0.9964)	1.028×10^{-5}
3.70×10^{-6} (−0.01)	−0.12253 (−4.06)			0.5	0.9985 (0.9966)	9.521×10^{-6}
0.0004368 (0.30)	−0.05875 (−4.14)			0.25	0.9985 (0.9968)	8.920×10^{-6}
0.005036 (0.59)	−0.02825 (−4.22)			0	0.9986 (0.9970)	8.362×10^{-6}
0.04240 (0.87)	−0.01367 (−4.29)			−0.25	0.9987 (0.9971)	7.872×10^{-6}
0.31428 (1.10)	−0.006649 (−4.29)	—	—	−5	0.9988 (0.9973)	7.471×10^{-6}
14.751 (1.50)	−0.001561 (−4.00)			−1	0.9988 (0.9974)	6.899×10^{-6}
651.22 (1.90)	−0.0003478 (−3.55)			−1.5	0.9989 (0.9976)	6.342×10^{-6}
4257.5 (2.13)	−0.0001617 (−3.42)			−1.75	0.9990 (0.9977)	5.973×10^{-6}
0^a (0)	-9.017×10^{-5} (−3.17)			−2.0	0.9979 (0.9962)	1.198×10^{-5}

The estimated l that maximizes the R^2 is −1.75 as opposed to the −0.5 for electrically heated dwellings. There are other similarities and differences between these results and those we derived for electrically heated homes. We note the following.

1. Both gas price vectors have a positive and statistically significant effect on insulation. Moreover, the boost in price responsiveness during the post-oil-crisis period is somewhat smaller than the extant responsiveness to prices that prevailed prior to 1975.

TABLE XIV

Average Short-Run Price Elasticity of Insulation Demand for Gas-Heated Homes, 1969–1980

Elasticities	Summer gas-heated dwellings	Summer electrically heated dwellings
Short-run energy price elasticity	0.028	0.008
Long-run energy price elasticity	0.061	0.070
Short-run income elasticity	−0.015	−0.034
Long-run income elasticity	−0.034	−0.290
Short-run energy-use elasticity	0.028	0.033
Long-run energy-use elasticity	0.064	0.280
Short-run interest elasticity	−0.028	−0.098
Long-run interest elasticity	−0.061	−0.822

Table XIV reports the summary results of our elasticity estimates. Figure 5 plots the elasticities for each year covered in the sample period.

During the post-oil-crisis period, the short-run price elasticity for gas-heated homes (with respect to gas price) was three times as large as it was for electrically heated dwellings (with respect to electricity price). For the long run, however, the price elasticities for gas- and electrically heated homes are about the same.

2. The estimated speed of adjustment for customers with gas-heated homes to changes in the economic environment is larger than the corresponding speed for customers with electrically heated homes. Perhaps this simply reflects the fact that both groups operate under an overall constraint on the annual increase in the insulation level, possibly because of short-run limitations on the capacity of the insulation industry in Sacramento. To the extent that the discrepancy between the target insulation level and the actual insulation level is larger for electrically heated than for gas-heated homes, a smaller fraction of the tension between the target and the existing levels of insulation will be made up for by electrically heated than by gas-heated homes. A cursory inspection of the year-to-year change in the insulation level suggests that this may at least partly explain the difference in the estimated speed of adjustment. But there may be other reasons. We did not investigate beyond this point.

As it is, equation 9* in Table XIII yields a speed of adjustment

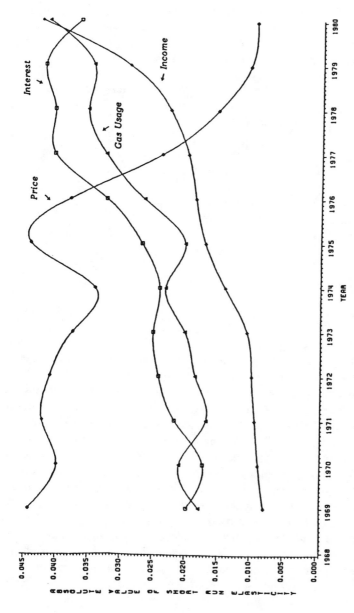

Fig. 5. Short-run elasticity of the residential demand for insulation of gas-heated homes (SMUD's service area, 1968–1980).

of 0.46 (= 1 − 0.54) for gas-heated homes, which simply says that approximately half of the total adjustment materializes in a little over a year. This is what the calculated mean lag of 1.2 years implies. [The estimate of the median lag yields an underestimate of the time it takes to complete 50% of the adjustment. Setting $1 − (0.54223)^{t+1} = 0.5$ yields $t = 0.13$ years, which becomes $t = 0.63$ years after adjusting to maintain equality of mean and median for a symmetric distribution.]

3. As in the case for electrically heated homes, insulation is an inferior good. However, both the short-run and long-run income elasticities are smaller for gas-heated than for electrically heated dwellings.

4. Elasticity with respect to energy consumption per household is twice as large as income elasticity. This is a reversal of what we found in the case of electrically heated homes in the winter. It also suggests that the net direct and indirect income effect on insulation may not be negative; alternatively, it may be negative, but only weakly so. (However, see the precautionary word at the end of this subsection.)

5. The interest rate variable has a strong effect. As in the case of electrically heated homes, its effect is always negative and statistically significant, regardless of the magnitude of the transformation parameter. But unlike the case of electrically heated homes, the interest rate is not the dominant variable.

6. Figure 5 shows that during the sample period, the insulation demand was in the process of becoming less and less sensitive to energy price, perhaps because even though the real gas price has been rising, it had been doing so only at a controlled rate. That is, the increase in real gas price may have been too moderate to make an impact. But even so, it is still as strong as the effect of electricity price on insulation demand for electrically heated homes, and it is probably the single most important factor that contributed to the increased insulation demand in the mid 1970s.

In contradistinction to its price elasticity, insulation demand has become increasingly sensitive to all three other determinants that appear in the model: income, interest rates, and energy usage. If the interest rates continue to decline (or as long as PG&E continues to offer the ZIP program) and if this is coupled with a sluggish economy, the recent upward trend in the insulation demand will probably continue.

Finally, a word of caution is in order. The results for the gas-heated model should be used with a great deal of caution, much more so than for electrically heated homes, because of the limited degrees of freedom for the gas-heated dwellings—only 12 observations. The model worked surprisingly well even though seven parameters (including the transformation parameters) were estimated. But we can see the signs of stress appearing when $l = -2$ was used in estimation. We did not pursue the estimation for gas-heated homes any further.

C. IOLS Results with Marginal Variances

We report in this subsection[20] the results of using the marginal variance calculated from the asymptotic covariance matrix of model 6* in Table XI. To make the transition from the conditional to the marginal variance, we focused on the correction term $FG^{-1}F'$ in (35). We calculated this term in two ways:

1. We used the inverse of the finite sample information matrix, i.e., the observed $(-\partial^2 L/\partial\theta, \partial\theta')^{-1}$. This is the approach proposed by Spitzer on the basis of Goldfeld and Quandt's work. The approach is motivated by the difficulty in evaluating analytically the expectations in $I(\theta) = [-E(\partial^2 L/\partial\theta \, \partial\theta')]$, where $I(\theta)$ is the information matrix. Spitzer's proposal is to drop the expectations operator in estimating the information matrix.

2. We used the inverse of the information matrix $I(\theta)$, which we calculated by numerical integration (Gauss Hermite quadrature available at U.C. Berkeley's computing center). We refer to these alternative calculations as finite sample and numerical integration, respectively. Table XV shows the results of calculating the correction terms $\sigma^2 FG^{-1}F'$.

Contrary to expectations, all estimates of the finite-sample correction terms turned out to be negative. (In the case of lagged dependent and use variables, the marginal variances themselves, as calculated by using Spitzer's approximation, were actually negative.) The correction

---

[20] This is a technical discussion. The reader may turn to Section VI.B for the result of using the limited information maximum likelihood (LIML) procedure to estimate the model. The LIML is a more appropriate procedure to use than the OLS when the dependent variable (insulation demand) is jointly determined with one or more of the explanatory variables (electricity usage).

TABLE XV

Estimates of the Correction Terms $\sigma^2 FG^{-1}F'$ for Model 6* of Table XI

Variable	Finite sample (Spitzer's approximation)	Numerical integration
Lagged dependent	−0.248011D − 02	0.567097D − 04
Income	−0.628589D − 02	0.159425D + 00
Interest rate	−0.488761D − 05	0.132336D − 02
Electricity price	−0.227831D − 08	0.240433D − 05
Electricity use	−0.3378760 + 00	0.506542D + 02

terms derived by numerical integration are all positive, as they should be.

The implications these corrections have for the estimates can be seen from Table XVI. For ease of comparison, the table shows also the *t* ratios of the IOLS estimates based on the conditional variance.

The column labeled finite sample shows a higher (we expect a lower) *t* ratio than the IOLS's *t*. As the table indicates, two of the *t* ratios were not calculated because the corrected variances were negative. The correction using numerical integration yields much lower *t*'s compared to the IOLS's *t*, with the one exception being the *t* ratio of the coefficient of the lagged variable.

The estimates were computed with 10-digit accuracy for IOLS, 16-digit accuracy for finite sample estimates, and 7-digit accuracy for numerical integration. We suspected that the finite sample results may have been influenced by round-off errors, perhaps due to differences in the magnitudes of the variables that enter the design matrix. (Prior to

TABLE XVI

_t_ Ratios of the Parameter Estimates Implicit in the Results of Table XV

Variable	IOLS	Finite sample (Spitzer's approximation)[a]	Numerical integration
Lagged dependent	15.60	NC	15.42
Income	−1.93	−2.81	−0.51
Interest rate	−4.02	−4.81	−0.44
Electricity price	3.11	3.23	0.35
Electricity use	2.39	NC	0.19

[a] NC means not calculated (negative variance).

that, we suspected that our programs or formulas may have been incorrect. We did extensive checking but found no error.) We considered standardizing by subtracting the mean and dividing by the standard deviation of each vector. But this standardization yields negative elements, which rules out estimating l on a continuous scale.

An alternative standardization procedure is to divide or multiply each vector of explanatory variables by an appropriate scalar so that the scale of our design matrix ranges between 0 and 10. Let the original model be

$$\left(\frac{Y_t}{\dot{Y}}\right)^{(l)} = b_1 + \sum_2^p b_j \left(\frac{X_{tj}}{\dot{Y}}\right)^{(l)} + u_t, \tag{48}$$

where $\dot{Y}$ is the geometric mean of Y_t. Scale the design matrix by dividing the jth vector by f_j,

$$\left(\frac{Y_t}{\dot{Y}}\right)^{(l)} = b_1^* + \sum_2^p b_j^* \left(\frac{X_{tj}}{\dot{Y}f_j}\right)^{(l)} + u_t^*, \tag{49}$$

Hence

$$b_j \left(\frac{X_{tj}}{\dot{Y}}\right)^l = b_j^* \left(\frac{X_{tj}}{\dot{Y}f_j}\right)^l \qquad \left(j = 2, \ldots, p,\right)$$

$$\Rightarrow \quad b_j = \frac{b_j^*}{f_j^l} \quad \text{and} \quad \hat{b}_j = \frac{\hat{b}_j^*}{f_j^l}. \tag{50}$$

When l is known,

$$\text{Var } \hat{b}_j = \text{Var } \hat{b}_j^* / f_j^{2l}. \tag{51}$$

When l is unknown,

$$\text{Avar}(\hat{b}_j) = \frac{1}{f_j^{2l}} [\text{Avar}(\hat{b}_j^*) + (\hat{b}_j^* \ln f_j)^2 \, \text{Avar}(\hat{l}) \\ -2(\hat{b}_j^* \ln f_j) \, \text{Acov}(\hat{b}_j^*, \hat{l})] \tag{52}$$

Equation (52) can be derived by taking the differential of the logarithm of (50), squaring both sides, and taking expectations.

We used (50) and (52) to back-transform to the original parameters. Table XVII shows the new correction terms. The story is the same as before. All finite sample correction terms are negative. The corrections derived by numerical integration are all positive. The implications of the results for the t ratio of the estimated parameters are shown in

TABLE XVII

Estimates of the Correction Terms $\sigma^2 F G^{-1} F'$ for Model 6* of Table XI[a]

Variable	Finite sample (Spitzer's approximation)	Numerical integration
Lagged dependent	−0.226153D − 03	0.423808D − 04
Income	−0.444913D − 03	0.199108D − 02
Interest rate	−0.215092D − 02	0.188376D − 01
Electricity price	−0.909759D − 05	0.552483D − 06
Electricity use	−0.144256D − 03	0.154439D − 02

[a] Data standardized to a scale of 0 to 10.

Table XVIII. Four of the coefficients have now negative variances. The coefficient of the lagged dependent variable which had a negative variance (see Table XVI) now has a positive variance, but one which is smaller than the IOLS's. Consequently, the t ratio is now higher than the IOLS's t.

The calculations in this round were carried out with 13-digit accuracy (versus 10 in Tables XV and XVI) for IOLS, 16-digit accuracy (as in Tables XV and XVI) for finite sample estimates; and 14-digit accuracy (versus 7 in Tables XV and XVI) for the numerical integration estimates.

Table XIX reports the results of estimating with the data of the design matrix standardized to a scale of 0 to 1. The computations are

TABLE XVIII

t Ratios of the Parameter Estimates Implicit in the Results of Table XVII

Variable	IOLS[a]	Finite sample (Spitzer's approximation)[b]	Numerical integration
Lagged dependent	13.79	14.41	13.69
Income	−2.10	NC	−0.46
Interest rate	−3.71	NC	−0.42
Electricity price	1.86	NC	0.21
Electricity use	2.39	NC	0.18

[a] The parameters and the t ratios are different from those reported in Tables XI and XVI reflecting the different levels of accuracy of the computations (see the text).

[b] NC means not calculated (negative variance).

TABLE XIX

Estimates of the Correction Terms $\sigma^2 F G^{-1} F'$ for Model 6* of Table XI[a]

Variable	Finite sample (Spitzer's approximation)	Numerical integration
Lagged dependent	0.151869D − 03	0.488840D − 06
Income	0.450121D − 05	0.203013D − 03
Interest rate	0.508401D − 04	0.100250D − 02
Electricity price	0.185756D − 05	0.192657D − 04
Electricity use	0.315657D − 05	0.945031D − 04

[a] Data standardized to a scale of 0 to 1.

done with a higher accuracy level for IOLS (14 digits) and for the numerical integration estimates (15 digits). The accuracy of the finite sample estimates is the same (16 digits).

The finite sample correction terms turned out positive this time, as did the correction terms derived by numerical integration, as they did before. But the results of Spitzer's approximation are still very different from those derived by numerical integration. For the lagged dependent variable, column 1 is more than 300 times larger than column 2. For income, the reverse is true: column 2 is 45 times as large as column 1. The same is true for the interest, price, and use, for which the multiples are 20, 10, and 30, respectively. The implication of the results for the t ratio of the estimates is shown in Table XX.

The t ratio according to the finite sample estimates is lower than its IOLS counterpart. But except for the lagged dependent variable, the t ratio of the estimate according to the finite sample estimates is two to

TABLE XX

t Ratios of the Parameter Estimates Implicit in the Results of Table XIX

Variable	IOLS	Finite sample (Spitzer's approximation	Numerical integration
Lagged dependent	15.25	14.82	15.25
Income	−1.52	−1.02	−0.52
Interest rate	−3.70	−1.57	−0.50
Electricity price	3.02	0.96	0.32
Electricity use	2.52	0.61	0.21

three times as large as for the t ratio reported under numerical integration.

Since the problem which led to the unexpected negativity we encountered seemed to afflict only the approximation proposed by Spitzer, we very much suspected that the problem may be due to the fact that Spitzer's approximation ignores the integrations altogether, even those that can be done analytically. Recall that Spitzer's approximation was motivated by the difficulty in evaluating analytically the expectation of the Hessian of L. However, there are terms in this matrix (specifically in the $E\partial^2 L^*/\partial l\ \partial a_k$) whose expectations are known analytically to be zero. We felt that this *a priori* information should be incorporated in Spitzer's approximation even though the rest of the expectations cannot be evaluated analytically. We did that and constrained these terms to be zero. Estimating this constrained version of Spitzer's approximation, we found out that all the negative signs of the correction terms were reversed: the terms became nonnegative, as they should be. Moreover, they turned out to be close to those we derived by numerical integration. The following is a brief rundown. Consider equation (A.6) in Appendix A. We have

$$E\left(-\frac{\partial^2 L}{\partial l\ \partial a_k}\right) = \sigma^{-2}E\left\{\sum_{t=1}^{T}\left(Y_t^{(l)} - \sum_{j=1}^{P}X_{tj}^{(l)}a_j\right)\left(-\frac{\partial X_{tk}^{(l)}}{\partial l}\right)\right.$$

$$\left. + \sum_{t=1}^{T}\left(\frac{\partial Y_t^{(l)}}{\partial l} - \sum_{j=1}^{p}\frac{\partial X_{tj}^{(l)}}{\partial l}a_j\right)(-X_{tk}^{(l)})\right\}. \qquad (53)$$

Calculating $\partial X_{tk}^{(l)}/\partial l$ and making use of (25), we can rewrite the first term on the right-hand side of (53) as

$$\sigma^{-2}E\left[\sum_{t=1}^{T}u_\tau\left(\frac{X_{tk}^l\ln X_{tk}^l}{l^2}\ \frac{X_{tk}^{(l)}}{l}\right)\right]. \qquad (54)$$

On the assumption that $X^{(l)}$ and u are contemporaneously uncorrelated, (54) and hence the first term on the right-hand side of (53) vanish.

Suppose, however, that we follow Spitzer's proposal and approximate by the observed Hessian [i.e., drop the expectations operator from (53)], and suppose that we estimate the parameters by using the maximum likelihood method. Denote by $\bar{u}$ the ML residual. Then (54) is replaced by

$$\sum_t\frac{X_{tk}^l\ln X_{tk}^l}{\bar{l}^2} - \sum_t\bar{u}_t\frac{X_{tk}^{(l)}}{\bar{l}}. \qquad (55)$$

The second term in (55) vanishes by the orthogonality of the explanatory matrix $X^{(i)}$ and the residuals vector $\bar{u}$. But the first term of (55) will not vanish. (The same thing, of course, happens when the parameters are estimated by using IOLS.) Since we know analytically that (54) vanishes, we can incorporate this information and constrain (55) to be zero. We did that. We found that Spitzer's approximation became well behaved.

Table XXI shows that whether we use the raw data matrix or scale the data to lie between 0 and 10 or between 0 and 1, the correction terms of the constrained version of Spitzer's approximation are all positive. Moreover, these terms are generally close to the correction terms derived by numerical integration (compare the results in columns 1, 2, and 3 of Table XXI with the last column of Tables XV, XVII, and XIX, respectively). Both estimates, the constrained Spitzer approximation and the numerical integration estimates, yield almost the same t ratios.

In light of these results and in light of the desirability in general of incorporating *a priori* information to improve the efficiency of the estimates, it is appropriate when using Spitzer's approximation of the marginal variance to constrain the terms in (55) to be zero rather than to estimate them from the sample. However, the question that logically precedes the issue of how to estimate the marginal variance is How meaningful are the estimates based on the marginal variance? In Section IV.A we pointed out that aside from the problem of sample size, the heterogeneity of the parameter for which the marginal variance is calculated raises a logical question about the meaningfulness of the marginal variance. The reason is that the marginal variance measures

TABLE XXI

Estimates of the Correction Terms $\sigma^2 F G^{-1} F'$ Derived from the Constrained Version of Spitzer's Approximation

Variable	Unscaled data	Data scaled to lie between	
		0–10	0–1
Lagged dependent	0.582326D − 04	0.592161D − 04	0.741769D − 07
Income	0.164220D + 00	0.185052D − 02	0.216691D − 03
Interest rate	0.1323280D − 02	0.156098D − 01	0.117104D − 02
Electricity price	0.235712D − 05	0.927967D − 06	0.211919D − 04
Electricity use	0.503611D + 02	0.130830D − 02	0.105017D − 03

the variance of magnitudes that are not comparable to begin with. The estimates we reported in this section cast further doubt on the reasonableness of results based on the marginal variance. Column 3 of Table XVI (and similarly column 3 of Tables XVIII and XX) suggests that electricity price, electricity usage, interest, and income, which the IOLS conditional variance showed to be significant, have in fact a statistically insignificant effect. It is difficult to reconcile these results with *a priori* expectations, since they are so counterintuitive. Moreover, they conflict the utilities' follow-up survey results on household motivation for conservation.

It is probably not unreasonable to conclude that the use of the conditional variance in inference, in spite of its limitations, probably does less violence to statistical inference than does the use of the marginal variance. At any rate, it may be worthwhile to take seriously the suggestion of Box and Cox (1982) and direct our efforts instead to an understanding of the circumstances under which estimates derived with a preassigned *l* provide an adequate basis for statistical inference.

Before we turn to LIML estimation, we shall report the result of one more avenue we traveled with an alternative procedure—the bootstrap method—in order to assess the variability of the estimates reported for model 6*, Table XI. As we shall see, the bootstrap portrays a picture close to that reported in Table XI for model 6*. The effort to derive estimates of the marginal variances for the bootstrap estimates did not take us further than the effort on which we reported in this section.

D. Bootstrap Estimation Results[21]

The bootstrap is a statistical procedure based on a resampling plan. It makes use of the data on hand only and permits an assessment of the variability (and other sampling properties) of the estimated parameters. The procedure approximates the underlying distribution of the disturbances by the empirical distribution of the observed residuals. By an appropriate resampling of the original observations, it constructs pseudodata on which the estimator is exercised. New parameter esti-

---

[21] This is a technical discussion. The reader may turn to Section VI.B for the results of using the limited information maximum likelihood (LIML) procedure to estimate the model. The LIML is a more appropriate procedure to use than the OLS when the dependent variable (insulation demand) is jointly determined with one or more of the explanatory variables (electricity usage).

mates are calculated every time a new set of pseudodata is generated. The variability of the new estimated parameters is used to assess the finite-sample behavior of the estimates derived from the original observations. The bootstrap is particularly helpful when the available results on estimate variability hold only asymptotically.

More generally, the procedure can be applied to much more general problems than assessing variability, particularly where parametric modeling or theoretical analysis may be too difficult for standardized analysis. Efron (1982) discussed the properties of the bootstrap and related resampling procedures. Freedman (1981) compared the bootstrap approximation with standard asymptotics. (For examples of applications to regression and econometric models using the distribution-free form, see Freedman and Peters, 1982, 1983.)

The bootstrap will not shed new light on the variability of the OLS estimates from a standard linear model, since it can be shown to give the standard estimate of covariance in the linear regression case (except for the estimate of the disturbance variance). The proof is straightforward (see Efron, 1982, p. 36). But no small sample results are available on how the variability of OLS and bootstrap estimates compare to each other when the lagged variable enters the model.[22]

In its simplest form the bootstrap is distribution free. But the procedure can also be executed in a parametric form by using exactly the same algorithm as for the distribution-free case, except that instead we can start with, say, a normal or a normal-smoothed or a uniform-smoothed empirical distribution.

We estimated most of our runs by using a normal empirical distribution. (If it is known that the disturbance term follows a normal distribution, one may expect the parametric results to be more efficient than those based on a distribution-free form.) A normal-probability plot (see Fig. 6) shows that the residuals from our model are well approximated by the normal distribution. Table XXII reports the bootstrap estimates using a normal empirical distribution and a distribution-free form for the residuals, all estimated on a scale of $\hat{l} = -0.5$.

To derive the bootstrap estimates, we resample with replacement 24 numbers from the residuals of our IOLS model 6* of Table XI. (Recall that the sample period has 24 observations.) We construct pseudodata

---

[22] Or when one or more explanatory variables are jointly determined with the dependent variable. But to use the bootstrap for this aspect one needs to compare the bootstrap results against the results of a consistent estimation method.

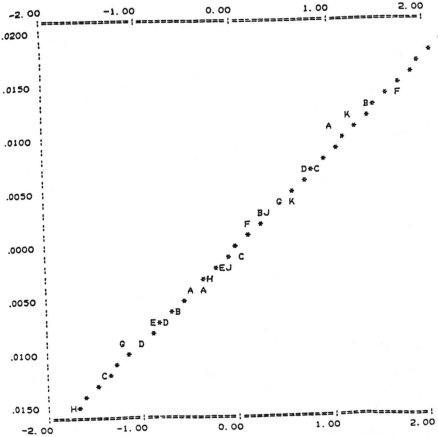

Fig. 6. Normal probability plot of the residuals of model 6*, Table XI.

by using these 24 drawings and use these pseudodata to estimate a new set of parameters. We repeat this process 100 times and use the standard deviation of the 100 estimates of each parameter as a measure of the variability of the bootstrap estimates. To be specific:

1. For each residual drawn, say, e_1^*, we construct a value for the dependent variable, call it $Y_1^{(l)*}$, by using the observations on the six transformed explanatory variables in period 1 together with the IOLS estimated parameters for each one of these variables as

TABLE XXII

IOLS and Bootstrap Estimates for Model 6* in Table XI[a]

	Intercept	Lagged insulation	Real income per HH	Real interest rate	Real electricity price	Electricity used per HH
Coefficient						
1. IOLS	−2.85311	0.88039	−0.24412	−0.01874	0.00064	1.59429
2. Bootstrap, normal	−3.18103	0.86118	−0.28130	−0.01905	0.00062	1.79079
3. Bootstrap, distribution free	−3.04362	0.86652	−0.25534	−0.01969	0.00061	1.69590
Standard error						
4. IOLS	1.18927	0.05643	0.12631	0.00466	0.00021	0.66620
5. Bootstrap, normal	1.40502	0.06366	0.14500	0.00499	0.00022	0.78686
6. Bootstrap, distribution free	1.22666	0.06188	0.11940	0.00420	0.00019	0.68322
Calculated t ratio						
7. 1:4	−2.399	15.599	−1.933	−4.024	3.048	2.393
8. 1:5	−2.031	13.830	−1.684	−3.756	2.909	2.026
9. 1:6	−2.326	14.227	−2.045	−4.462	3.368	2.222
RMS error						
10. Bootstrap, normal	1.29824	0.06339	0.13683	0.00491	0.00021	0.73020
11. Bootstrap, distribution free	1.07544	0.05375	0.11690	0.00414	0.00018	0.60775
Ratio of bootstrap standard deviation to RMS						
12. 10:5	0.924	0.996	0.944	0.984	0.955	0.928
13. 11:6	0.877	0.869	0.979	0.986	0.947	0.890
Tests for significant difference						
14. $\dfrac{10*(R1 - R2)}{R5}$	2.33	3.02	2.56	1.56	−0.91	−2.50
15. $\dfrac{10*(R1 - R3)}{R6}$	1.55	2.24	0.94	2.26	1.58	−1.49

[a] $l = 0.5$; 100 replicates.

reported in 6* Table XI. Denoting the transformed explanatory variables for period 1 by $X_{i1}^{(i)*}$, we have

$$Y_1^{(i)*} = \sum_{i=1}^{6} b_i X_{i1}^{(i)*} + e_1^*,$$ (56)

where the six b_i's are listed in equation 6* of Table XI. Note that the row vector $X_{.t}^{(i)*}$ is identical to the row vector $X_{.t}^{(i)}$ used in the IOLS estimation.

2. Replacing e_1^*, we make another drawing from the 24 residuals of our IOLS model. The new drawing e_2^* is used to construct $Y_2^{(i)*}$ in the same manner as in (56), with the six transformed values $X_{i2}^{(i)*}$ and the same b_i's used as before. However, because a lagged value of Y constitutes one of the columns of the X matrix, the construction of $Y_t^{(i)*}$ for $t = 2, \ldots, 24$ proceeds iteratively with $Y_{t-1}^{(i)*}$ entering the construction of the row vector $X_{.t}^{(i)*}$. This process of drawing e_t^* with a replacement and constructing the pseudodata is repeated until we have constructed 24 $Y_t^{(i)*}$'s.

3. We regress these 24 values of $Y_{it}^{(i)*}$ on the $X_{it}^{(i)*}$'s and derive our first batch of new estimates of $b_{i(1)}^*$, $i = 1, \ldots, 6$. We repeat steps 1–3 to generate a second set of estimates $b_{i(2)}^*$, using our second set of 24 manufactured $Y_t^{(i)*}$'s, and so on. We may replicate this experiment until we have generated 50 or 100 such sets of estimates. Table XXII reports the results we derived based on 100 replications separately for normal and distribution-free forms of the residuals. All bootstrap computations were performed by using a program originally available from the TROLL Econometric Modeling System and run on IBM 4341 available at U.C. Berkeley's computing center.

For ease of reference, row 1 of Table XXII shows the IOLS results we reported in 6*, Table XI. The bootstrap estimates are shown in rows 2 and 3. Each estimate shown in rows 2 and 3 is the average value of the coefficients calculated in 100 replications.

A comparison of row 1 against rows 2 and 3 shows that, perhaps with the exception of the coefficients of the intercept and the electricity use per household, both the normal and distribution-free forms yield estimates that are *numerically* very close to the IOLS estimates.

Row 4 shows the standard errors calculated from the IOLS. Rows 5 and 6 report the standard deviation of the 100 replicates of the estimated parameters. The values in these two rows may be viewed as a measure of the "true" variability of the procedure of estimating the bootstrap coefficients.[23]

The *t* ratios calculated from the IOLS estimates are shown in row 7. One indicator of the reliability of the ratios reported in this row is listed in row 8 for the normal case and row 9 for the distribution-free case. Here we calculate the ratio of the IOLS estimated parameter to the bootstrap standard deviation. The assumption underlying these calculations is that the results derived in a bootstrap world carry over to the real world of the households—which may or may not be the case. For ease of reference, rows 7–9 of Table XXII are reproduced (to two decimals) in the accompanying tabulation.

	IOLS estimate: IOLS standard error (row 7)	IOLS estimate: bootstrap normal standard deviation (row 8)	IOLS estimate: bootstrap distribution-free standard deviation (row 9)
Intercept	−2.40	−2.03	−2.33
Lagged insulation	15.60	13.83	14.23
Real income/HH	−1.93	−1.68	−2.05
Real interest rate	−4.02	−3.76	−4.46
Real electricity price	3.05	2.91	3.37
Electricity use/HH	2.39	2.03	2.22

There is an indication of a downward bias in the estimate of the variability of the IOLS parameters. On the other hand, the *t* ratios in the second and third columns may themselves be underestimates. This

---

[23] The standard deviation of the bootstrap estimates in row 2 is $\frac{1}{10}$ standard deviation shown in row 5, since $\sigma_{\bar{x}} = \sigma_x/N^{1/2}$. Rows 3 and 6 are related similarly. We can use the results in rows 5 and 6 to test whether the estimates in rows 2 and 3 are statistically significantly different from the estimates in row 1, even though rows 1–3 are numerically close. The results of these calculations are listed in rows 14 and 15 of Table XXII. We do not make use of these results, since we are interested in the overall picture conveyed by the conditional results compared to our earlier IOLS conditional results.

is so because to estimate the variance of the disturbance for bootstrap purposes, we divided the sum of the squares of the residuals from the IOLS by $T - k$ rather than by T. [Efron (1982, p. 36) divides by T. This is as it should be since the residuals from the model are the population from which the drawing is made. The variance of the uniform distribution (or the normal distribution generated from the uniform) as used in simulation is simply the average of the sum of the squares of these residuals, where the average is derived by dividing by T. But as Efron notes (p. 36), we can get falsely optimistic results if we are fitting a highly overparameterized model. In that case, one may want to divide by $T - k$. But the general case is unclear.]

As it is, the normal bootstrap shows that the downward estimate of variability occurs in the case of every IOLS parameter estimate. However, the distribution-free results do not seem to confirm this general tendency. They show that IOLS overstates the variability of the estimates for income, price, and interest rate.

An alternative and perhaps more meaningful way of evaluating the reliability of the IOLS estimates of variability is to compare row 10 to row 5 and row 11 to row 6 in Table XXII. Row 10 is the root mean square (RMS) of the diagonal elements of the OLS covariance matrix for the vector $b^*_{.(n)}$, $n = 1, ..., 100$. The RMS provides a measure of the typical standard error for the parameters estimated from the pseudo-data $Y_t^{(i*)}$, using the OLS standard expression for the variance of the estimate. We would expect these RMS values to be about the same as the standard deviations shown in row 5. Similar remarks apply to the values reported in rows 11 and 6.

A comparison of row 5 to row 10 and row 6 to row 11 shows that (see rows 12 and 13 in Table XXII), in general, the OLS expression for the standard error of the estimates tends to be biased downward. (Again, had we calculated the variance by dividing by T, none of the downward biases would have appeared. In fact, we would have had to conclude that the IOLS standard error overstates the variability.)

As it is, rows 12 and 13 taken together suggest that the bias is largest in the case of the intercept and the consumption per household. For the normal case (which is the more relevant case since our residuals are approximately normal) the bias is relatively small. At any rate, even for the intercept and the electricity usage variable, the bias is smaller than 8% of the bootstrap standard deviation.

In summary, the story conveyed by the bootstrap results is consis-

TABLE XXIII

IOLS and Unconditional Bootstrap Estimates[a]

	Intercept	Lagged insulation	Real income per HH	Real interest rate	Real electricity price	Electricity used per HH	Transformation parameter
Coefficient							
1. IOLS	−2.85311	0.88039	−0.24412	−0.01874	0.00064	1.59429	−0.50
2. Bootstrap, normal	−18.09706	0.84585	−0.43387	−0.03630	0.00139	15.95472	−0.48
Standard error							
3. IOLS	1.18927	0.05643	0.12631	0.00466	0.00021	0.66620	
4. Bootstrap	27.03880	0.06354	0.48358	0.03285	0.00140	24.97082	0.32902
Calculated t ratio							
5. 1:3	−2.399	15.599	−1.933	−4.024	3.109	2.393	
6. 1:4	−0.106	13.856	−0.505	−0.570	0.457	0.064	1.520

[a] 50 replicates.

tent with the story conveyed by the IOLS results for model 6* shown in Table XI. Both procedures agree on the statistical significance of the explanatory variables we used in estimating our model. This is a rather important finding, since it is not known *a priori* how the variability of the bootstrap relates to the variability of the IOLS estimates when a lagged dependent variable appears in the system.

We went one step further and asked: What if $\hat{l}$ were not given a preassigned value. (Of course, the same objections discussed earlier regarding the use of the marginal variance are applicable here as well. But this time we were interested in finding out what mean value and variance the bootstrap will calculate for the MLE of l.) Table XXIII reports the results. To save on cost, we generated simulation results by using the empirical normal distribution only. We also constrained the grid over which the search for the MLE was conducted and confined ourselves to 50 replicates.

For the purpose of this experiment, the $Y_t^{(\hat{l}*)}$ and the matrix $X_t^{(\hat{l}*)}$ were generated exactly in the same manner as before by using $\hat{l} = -0.5$ and using the same b_i's listed in equation 6* of Table XI. But because this time we are interested in deriving the MLE of the transformation parameter, we carry out the following procedure.

We generate first a (24×1) vector $Y^{(l)*}$ and a (24×6) matrix $X^{(\hat{l}*)}$, where $\hat{l} = -0.5$. We now detransform to derive Y^* and X^* with elements

$$Y_t^* = (\hat{l} Y_t^{(\hat{l})*} + 1)^{1/\hat{l}}, \tag{57}$$

$$X_{it}^* = (\hat{l} X_{it}^{(\hat{l})*} + 1)^{1/\hat{l}}, \tag{58}$$

respectively. We postulate the model

$$Y^{*(l^*)} = X^{*(l^*)} b^* + u^*, \tag{59}$$

with

$$u^* \sim N(0, \sigma^{2*} I), \tag{60}$$

and normalize the variables in (59) by dividing each element in Y^* and X^* by the geometric average of the elements in Y^*, $\dot{Y}^*$.

To estimate $(\hat{l}_{(1)}^*, \hat{b}_{\cdot(1)}^*)$, where the subscripts (1) denote the estimate derived from the first replicate, we use IOLS and search over a grid for the $\hat{l}^*$ that yields the smallest sum of the squares of the residuals for the normalized version of (59). (When the grid includes 10 values of l^*, we generate 10 IOLS estimates and pick up the set that minimizes the sum of the squares of the errors.) Since the variables are all normalized, the vector that yields the smallest sum of the squares of the residuals is also the vector of MLE. We do the same thing with the second replicate to derive the MLE $(\hat{l}_{(2)}^*, \hat{b}_{\cdot(2)}^*)$, and so on, for each one of the 50 replicates of the experiments. Table XXIII shows the results.

Row 1 shows the IOLS results. Row 2 shows that the bootstrap estimates the transformation parameter at $\hat{l} = -0.48$, which is numerically close to, and statistically not different from, $l = -0.5[(0.50 - 0.48)/(0.32902/10) = 0.61]$. But we also note that the coefficients of the intercepts and the electricity usage are 6.5 and 10 times as large as the IOLS estimates. (Recall that in examining Table XXII, we noticed divergences between the IOLS and the bootstrap estimates for the parameters of these two variables.) The results of singular value decomposition indicated that a multicollinearity exists among the intercept, electricity usage, and lagged insulation, which concentrates on the first two variables. It is possible the bootstrap is too sensitive to problems of multicollinearity, and perhaps this aggravates the drop in the ratio of the estimate to its standard error.

Whatever the case may be, since we have no way of increasing our sample size to overcome this problem, we explored the impact of reducing the strain on the available degrees of freedom by reducing the number of estimated parameters. We dropped the electricity usage variable from the equation and did notice an improvement in the equation. But we did not continue our probing in this direction any further (nor did we look into the problem that some variables are numerically large in comparison to others), partly because of cost constraints and partly because we did not intend to use the marginal variance in our analysis in any case.

VI. CONSISTENT ESTIMATION

A. The Likelihood Function for the LIML of a Box–Cox-Transformed System[24]

We follow Dhrymes's (1970) framework for the derivation of LIML in the standard case, but extend it to a Box–Cox-transformed model. Consider the structural model

$$Z^{(l)}A = U, \tag{61}$$

where

$$Z^{(l)} = (Y^{(l)}, X^{(l)}), \tag{62}$$

$$A = \begin{bmatrix} I - B \\ -C \end{bmatrix}, \tag{63}$$

and $Y^{(l)}$ and $X^{(l)}$ are, respectively, the $T \times m$ and $T \times G$ matrices of transformed current endogenous and transformed predetermined variables, with typical elements

$$z_{ti}^{(l)} = (z_{ti}^l - 1)/l, \tag{64}$$

and B and C are, respectively, the $m \times m$ and $G \times m$ matrices of unknown parameters. The rows of U, $u_{t.}$ have the properties

$$u_{t.}' \sim N(0, \Sigma), \qquad E(u_{t.}' u_{s.}) = \delta_{ts}\Sigma \qquad (t, s = 1, 2, \ldots, T), \tag{65}$$

---

[24] This is a technical discussion. The reader may turn to Section VI.B for the estimation results.

266   J. Daniel Khazzoom

where δ_{ts} is the Kronecker delta and Σ a positive definite matrix. We also make the usual assumption that

$$\text{rank } X = G. \tag{66}$$

Suppose that we are interested in estimating the parameters of the first m^* equations in (61). We seek a transformation of the system, call it H, such that when we apply H to (61), the following two conditions are met:

1. The two subsystems consisting of the first m^* equations and the remaining $(m - m^*)$ equations become independent; i.e. the cross covariance between the two subsystems is the null matrix.
2. The parameters of the first m^* equations are not disturbed.

The tth row $u_t.H$ of the transformed error vector UH has the property

$$\text{Cov}(u_t.H, u_s.H) = \delta_{ts}H'\Sigma H \tag{67}$$

because of the independence of the vectors $u_t.$ and $u_s.$. The joint density of $u_t.H$ is

$$p(u_1.H, \ldots, u_T.H)$$
$$= (2\pi)^{-mT/2}|H'\Sigma H|^{-T/2}$$
$$\exp\left\{-\frac{1}{2}\sum_{t=1}^{T}(u_t.H)(H^{-1}\Sigma^{-1}H'^{-1})(u_t.H)'\right\}. \tag{68}$$

For any *fixed* t, consider the transformation

$$u_t.H = y_t^{(l)}(I - B)H - x_t^{(l)}CH \tag{69}$$

and note that the Jacobian from $u_t.H$ to $y_t.$ is given by

$$J = J_1 * J_2, \tag{70}$$

where

$$J_1 = J(u_{ti}H \rightarrow y_{ti}^{(l)}) = |(I - B)H|, \tag{71}$$

$$J_2 = J(y_{ti}^{(l)} \rightarrow y_{ti}) = \prod_{i=1}^{m} y_{ti}^{l-1}. \tag{72}$$

Hence the log-likelihood function of the current endogenous variables is (as t ranges from 1 to T)

$$L^*(A, H, \Sigma; Y, X) = -\frac{mT}{2}\ln(2\pi) - \frac{T}{2}\ln|H'\Sigma H|$$

$$- \frac{1}{2}\sum_{t=1}^{T}(z_{t.}^{(l)}AH)(H'\Sigma H)^{-1}(z_{t.}^{(l)}AH)' \qquad (73)$$

$$+ T\ln|(I - B)H| + (l - 1)\sum_{t=1}^{T}\sum_{i=1}^{m}\ln y_{ti},$$

where

$$z_t^{(l)} = (y_{t.}^{(l)}, x_{t.}^{(l)}). \qquad (74)$$

Any parameters of the subsystem of $(m - m^*)$ equations that remain in (73) can be eliminated by partially maximizing (73) with respect to these parameters and substituting their maximizing values back in (73). The resulting concentrated likelihood can then be maximized with respect to the parameters of the first m^* equations.

We are interested in the case in which $m^* = 1$. Specifically, suppose that we are interested in the first structural equation

$$y_{t1}^{(l)} = \sum_{j=2}^{m}\beta_{j1}y_{tj}^{(l)} + \sum_{s=1}^{G}\gamma_{s1}x_{ts}^{(l)} + u_{ti} \qquad (75)$$

with the understanding that only $m_1 + 1$ of the β's and G_1 of the γ's are known not to be zero.

Following Dhrymes's notations we employ the numbering convention that the variables actually appearing in the first equation are $y_1^{(l)}$, ..., $y_{m_1+1}^{(l)}$, $x_1^{(l)}$, ..., $x_{G_1}^{(l)}$, so that

$$Y_1^{(l)} = (y_{.1}^{(l)}, ..., y_{.m_1+1}^{(l)}), \qquad X_1^{(l)} = (x_{.1}^{(l)}, ..., x_{.G_1}^{(l)}) \qquad (76)$$

are the vectors of T observations on $y_i^{(l)}$ and $x_s^{(l)}$ that appear in the equation.

Similarly, let

$$\beta_{.1}^0 = (1 - \beta_{21} \cdots -\beta_{m_1+1,1})' = (1 - \beta_{.1})',$$

$$\gamma_{.1} = (\gamma_{11}\gamma_{21} \cdots \gamma_{G_11})' \qquad (77)$$

be, respectively, the $(m_1 + 1) \times 1$ and the $G_1 \times 1$ vectors of parameters that appear in the first structural equation. Define the moment matrices as

$$M^{(l)} = \frac{1}{T} \begin{bmatrix} Y^{(l)\prime} Y^{(l)} & Y^{(l)\prime} X^{(l)} \\ X^{(l)\prime} Y^{(l)} & X^{(l)\prime} X^{(l)} \end{bmatrix} = \begin{bmatrix} M_{yy}^{(l)} & M_{yx}^{(l)} \\ M_{xy}^{(l)} & M_{xx}^{(l)} \end{bmatrix}$$

$$= \begin{array}{c} \\ m_1+1 \\ m-(m_1+1) \\ G_1 \\ G-G_1 \end{array} \begin{array}{cccc} {\scriptstyle m_1+1} & {\scriptstyle m-(m_1+1)} & {\scriptstyle G_1} & {\scriptstyle G-G_1} \\ \begin{bmatrix} M_{11}^{(l)} & M_{12}^{(l)} & M_{13}^{(l)} & M_{14}^{(l)} \\ M_{21}^{(l)} & M_{22}^{(l)} & M_{23}^{(l)} & M_{24}^{(l)} \\ M_{31}^{(l)} & M_{32}^{(l)} & M_{33}^{(l)} & M_{34}^{(l)} \\ M_{41}^{(l)} & M_{42}^{(l)} & M_{43}^{(l)} & M_{44}^{(l)} \end{bmatrix} \end{array}, \qquad (78)$$

$$W^{(l)} = M_{yy}^{(l)} - M_{yx}^{(l)} M_{xx}^{(l)^{-1}} M_{xy}^{(l)}$$

$$= \begin{array}{c} \\ m_1+1 \\ m-(m_1+1) \end{array} \begin{array}{cc} {\scriptstyle m_1+1} & {\scriptstyle m-(m_1+1)} \\ \begin{bmatrix} W_{11}^{(l)} & W_{12}^{(l)} \\ W_{21}^{(l)} & W_{22}^{(l)} \end{bmatrix} \end{array}, \qquad (79)$$

with

$$W_{11}^{(l)} = M_{11}^{(l)} - [M_{13}^{(l)} \quad M_{14}^{(l)}] \begin{bmatrix} M_{33}^{(l)} & M_{34}^{(l)} \\ M_{43}^{(l)} & M_{44}^{(l)} \end{bmatrix}^{-1} \begin{bmatrix} M_{31}^{(l)} \\ M_{41}^{(l)} \end{bmatrix}. \qquad (80)$$

The LIML estimator of the parameters of the first structural equation can be derived from (73), which, following Dhrymes's equation (7.3.58), p. 337, becomes

$$L^*(\beta_{.1}^0, \gamma_{.1}, \sigma_{11}, l; Y; X)$$

$$= -\frac{mT}{2}(\ln(2\pi) + 1) + \frac{T}{2}(1 - \ln|W^{(l)}|)$$

$$- \frac{T}{2}\ln\sigma_{11} + \frac{T}{2}\ln(\beta_{.1}^{0\prime} W_{11}^{(l)} \beta_{.1}^0)$$

$$- \frac{T}{2}\frac{1}{\sigma_{11}}(\beta_{.1}^{0\prime} M_{11}^{(l)} \beta_{.1}^0 - 2\gamma_{.1}^\prime M_{31}^{(l)} \beta_{.1}^0 + \gamma_{.1}^\prime M_{33}^{(l)} \gamma_{.1})$$

$$+ (l - 1)\sum_{i=1}^{m}\sum_{t=1}^{T}\ln y_{ti}. \qquad (81)$$

The computation of the LIML estimates can be done by searching over a grid of values for l, as described earlier, and choosing that value of $\hat{l}$ that maximizes (81). But because of the way the SAS program we used solves for LIML, we went through two stages to solve for the maximum likelihood estimate of l. [We used the SAS program version 79.6 available on IBM 4341 at U.C. Berkeley's computing center. The program does not calculate the LIML estimates by maximizing the likelihood function (81). The estimate of $\beta^{0'}_{.1}$ is first calculated by solving for the minimum ratio $\beta^{0'}_{.1} W^{(l)*}_{11} \beta^{0}_{.1} / \beta^{0'}_{.1} W^{(l)}_{11} \beta^{0}_{.1}$, where $W^{(l)*}_{11}$ is the second moment matrix of residuals from the regression of Y_1 on X_1 and $W^{(l)}_{11}$ is as defined in (80)—namely, the second moment matrix of the residuals from the regression of Y_1 on X. $\beta^{0}_{.1}$ is then substituted in the expressions of the first-order condition to derive the LIML estimates of $\gamma_{.1}$ and σ_{11}.] Briefly, the program maximizes with respect to the parameters in (81) while lumping the term $-(T/2) \ln|W^{(l)}|$ in (81) with the rest of the constants. But if we are to maximize over l, we cannot treat the term $\ln|W^{(l)}|$ as if it remained constant when l varies. Hence the SAS procedure had to be modified.

To avoid rewriting a new program, we followed the following two-step procedure. We used the SAS LIML program to estimate $\beta^{0}_{.1}, \gamma_{.1}$, and σ_{11}, conditional on a stipulated l_k. (Note that for any given l, the term $\ln|W^{(l)}|$ is a constant.) Next, we concentrated the estimated parameters out of the likelihood function and evaluated the conditional log-likelihood $L^*(l_k)$ in (81). (Actually we ignored the constants $-[mT/2][\ln(2\pi + 1)] + T/2$. We repeated this process for other values of l_k, $k = 1, \ldots, 7$, and chose that $L^*(\hat{l}_k)$ that has the highest value. The results are shown in Table XXIV.

One observation: As before, we simplify the computations by normalizing the variables. We divide each z_{ti} by $\ddot{y} = (\prod_{t=1}^{T} \prod_{i=1}^{m} y_{ti})^{1/Tm}$ and use $z^{*(l)}_{ti} = [(z_{ti}/\ddot{y})^l - 1]/l$ instead of $z^{(l)}_{ti}$ in the estimation. When the variables are thus normalized, the term $(l - 1) \sum_{t=1}^{T} \sum_{i=1}^{m} \ln y_{ti}$ drops out of (81). By substituting $(z^{(l)}/\ddot{y}^l) - (\ddot{y}^{(l)}/\ddot{y}^l)$ for $z^{*(l)}$, it is straightforward to show that except for the intercept of the estimated equations, the rest of the parameters are the same as the parameters we obtain if we were to estimate without normalization.

The results reported in Table XXIV show the ratio of the LIML estimate to its asymptotic standard error, conditional on l. We need now to derive the marginal asymptotic covariance as well. Asymptotically, we have

$$(T)^{1/2} \begin{bmatrix} \hat{\beta}^0_{.1} - \beta^0_{.1} \\ \hat{\gamma}_{.1} - \gamma_{.1} \\ \hat{\sigma}_{11} - \sigma_{11} \\ \hat{l} - l \end{bmatrix} \xrightarrow{d} \left(0, \plim_{T \to \infty} \left(\frac{I_T}{T} \right)^{-1} \equiv \sigma_{11} S^{-1} \right), \tag{82}$$

TABLE XXIV

LIML Estimates of the Residential Demand for Insulation for All Electrically Heated Homes in SMUD's Service Area, 1969–1980

Equation	Coeff. and t ratio	Intercept (1)	Lagged insulation (2)	Real electricity price, 1974–1980 (3)	Real income per HH (4)	Use per HH (5)	Real interest rate (6)	Column with incorrect sign (7)	Correct sign but $t \leq 1$ (8)	l (9)	Value of the log-likelihood function
1	Coeff.	-213.133	0.921611	0.0000641	-0.753987	63.28675	-0.002865			-1	-530.15
	t	(-1.724)	(17.065)	(3.258)	(-1.065)	(1.666)	(-3.391)				
2.	Coeff.	-55.3975	0.909024	0.0002130	-0.417233	9.55678	-0.007279			-0.75	-485.57
	t	(-2.059)	(15.843)	(3.217)	(-1.371)	(1.823)	(-3.649)				
3	Coeff.	-16.0073	0.891777	0.0006197	-0.225970	1.44082	-0.018324			-0.5	-501.93
	t	(-2.471)	(15.032)	(3.144)	(-1.733)	(2.044)	(-3.878)				
4	Coeff.	-5.40182	0.872379	0.0013926	-0.117739	0.212711	-0.045414			-0.25	-516.82
	t	(-3.027)	(14.689)	(3.083)	(-2.106)	(2.326)	(-4.023)				
5*	Coeff.	-2.34254	0.857886	0.0018182	-0.057439	0.029887	-0.109637	—		0	-454.41
	t	(-3.567)	(14.428)	(3.075)	(-2.351)	(2.549)	(-4.018)				
6	Coeff.	-1.69000	0.816476	0.0010943	-0.031143	0.0049530	-0.272434			0.25	-701.11
	t	(-3.955)	(11.646)	(2.995)	(-2.615)	(2.809)	(-3.914)				
7	Coeff.	-1.66787	0.805306	0.0003946	-0.014383	0.0006809	-0.636639			0.5	-751.83
	t	(-4.068)	(12.017)	(2.945)	(-2.822)	(3.248)	(-3.651)				

where we define

$$S_{a,b} = \sigma_{11} \operatorname{plim}\left(-\frac{1}{T}\frac{\partial^2 L}{\partial a\ \partial b}\right) \qquad \text{for} \quad a, b = (\beta_{.1}^0, \gamma_{.1}, \sigma_{11}, l)$$

$$= \begin{bmatrix} S_{\beta_{.1}^0\beta_{.1}^0} & S_{\beta_{.1}^0\gamma_{.1}} & S_{\beta_{.1}^0\sigma_{11}} & S_{\beta_{.1}^0 l} \\ & S_{\gamma_{.1}\gamma_{.1}} & 0 & S_{\gamma_{.1}l} \\ & & S_{\sigma_{11}\sigma_{11}} & S_{\sigma_{11}l} \\ & & & S_{ll} \end{bmatrix} \qquad (83)$$

The derivation of I_T and S in (82) is shown in Appendix C. The results of recomputing the ratios of the LIML estimates to their asymptotic standard error, while treating l as an estimated parameter, are shown in Table XXVI. In computation we constrain the first element of $\beta_{.1}^0$ to be 1.

B. LIML Results with Conditional Variance for Electrically-Heated Homes

Table XXIV shows the LIML estimates for electrically heated homes in the same format explained in Section V.A. There are two endogenous variables in the equation: (1) current insulation level of electrically heated homes and (2) electricity used per electrically heated household. Excluding the intercept, there are seven predetermined variables, of which the first four appear in the equation: (1) lagged value of insulation level in electrically heated homes, (2) real income per household in Sacramento County, (3) real interest rate, (4) real price of electricity, (5) lagged value of the aggregate appliance efficiency index, (6) lagged value of the demand for electricity for electrically heated homes, and (7) lagged value of the number of households with electric heat. The table labels the ratio of the estimate to its asymptotic standard error t ratio. Asymptotically, the ratio is approximately normally distributed.

The table shows that $l = 0$ gives the highest value for the likelihood function. (See model 5*.) We did not search on a more finely divided grid. A plot of the estimated log likelihood, smoothed by a cubic spline (see Fig. 7), shows that the maximizing l value is very close to zero. Subsequent estimation suggests that l is tightly centered around zero, with a variance practically equal to zero. See Section VI.C.

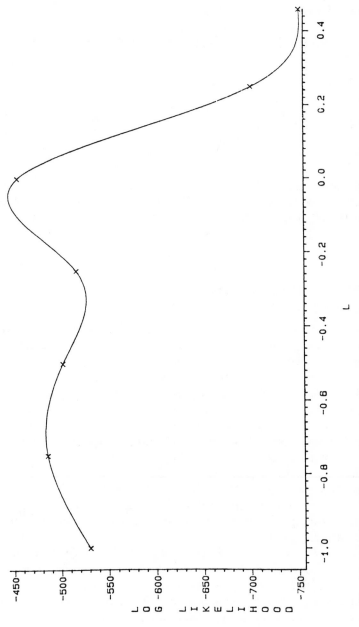

Fig. 7. Value of the log-likelihood function for the conditional LIML estimates of the residential demand for insulation by all electrically heated homes (SMUD's service area, 1969–1980).

In general, the LIML estimates are close to their IOLS counterparts except for the intercept (which swings much more widely with the IOLS than it does with the LIML) and the estimate for the price of electricity. Otherwise, all estimates have the right sign with the ratio of the parameter to its (conditional) asymptotic standard error, generally exceeding 2. These ratios may be conservative, since we have chosen to estimate the variance of the disturbance term by dividing by $T - k$ rather than T as the *ML* requires. Had we divided by T instead, the ratios of the estimates to their asymptotic standard errors would have been higher by 13%.

It is not possible to undertake an analysis of the elasticities and other measures implicit in the LIML estimates. A full analysis requires a solved reduced form. We did note earlier that the LIML estimates, except for the price coefficient (and intercept), are generally close to the IOLS estimates. Other than that, however, and without allowing for the feedback effect, it appears that the short-run interest rate elasticity implicit in the LIML estimates is about the same as for the one-equation model. It is similarly the largest of all other short-run elasticities. The short-run elasticity with respect to electricity price ranks lowest, as was the case before, but is less than one-tenth of the corresponding summer and winter price elasticities estimated by the one-equation model. (In practice, the electricity price elasticity is likely to be even smaller than that suggested by the LIML estimates because of the depressing effect of the higher electricity price on electricity demand.) The short-run income elasticity is somewhat larger than the corresponding summer and winter income elasticities implicit in the one-equation model. (The income elasticity is likely to be smaller than that suggested by the LIML estimates because of the positive effect of income on the demand for electricity.) The short-run electricity-usage elasticity falls somewhere between the summer and winter elasticities implicit in the one-equation model.

Table XXV reports the two-stage least-squares (2SLS) estimates.[25] We note that for the same *l*, these estimates tend to lie between the IOLS and the LIML estimates. In general we do not know that these estimates are consistent. But we will report in Section VI.E nonlinear two-stage least-squares estimates. These are expected to be consistent

---

[25] R^2 does not have a lower bound when calculated for equations estimated by 2SLS or other limited information methods. It is reported in Table XXV only as a rough indicator for the seven sets of transformed equations we estimated.

TABLE XXV

2SLS Estimates of the Residential Demand for Insulation by All Electrically Heated Homes in SMUD Service Area, 1969–1980

Equation	Coeff. and t ratio	Intercept (1)	Lagged insulation (2)	Real electricity price, 1974–1980 (3)	Real income per HH (4)	Use per HH (5)	Real interest rate (6)	Column with incorrect sign (7)	Correct sign but $t \leq 1$ (8)	l	R^2
1	Coeff	-233.642	0.917117	0.0000637	-0.798879	66.6568	-0.0029056	—	—	-1	0.9973
	t	(-1.830)	(17.173)	(3.250)	(-1.138)	(1.777)	(-3.459)				
2	Coeff.	-57.1687	0.905920	0.0002117	-0.432472	9.91294	-0.007348	—	—	-0.75	0.9974
	t	(-2.126)	(15.939)	(3.206)	(-1.431)	(1.912)	(-3.699)				
3	Coeff.	-16.2402	0.889643	0.0006176	-0.229613	1.46530	-0.018409	—	—	-0.5	0.9974
	t	(-2.528)	(15.126)	(3.137)	(-1.771)	(2.101)	(-3.906)				
4	Coeff.	-5.40495	0.872272	0.0013925	-0.111782	0.212881	-0.045423	—	—	-0.25	0.9974
	t	(-3.051)	(14.800)	(3.084)	(-2.12)	(2.348)	(-4.030)				
5	Coeff.	-2.31948	0.860108	0.0018214	-0.056746	0.029431	-0.109217	—	—	0	0.9973
	t	(-3.572)	(14.648)	(3.082)	(-2.339)	(2.544)	(-4.012)				
6	Coeff.	-1.58162	0.837357	0.0010985	-0.028310	0.0044102	-0.263857	—	—	0.25	0.9971
	t	(-3.954)	(13.000)	(3.069)	(-2.525)	(2.736)	(-3.909)				
7	Coeff.	-1.64498	0.811373	0.0003939	-0.014027	0.0006611	-0.631367	—	—	0.5	0.9966
	t	(-4.068)	(12.345)	(2.969)	(-2.795)	(3.2175)	(-3.661)				

TABLE XXVI

LIML Results with Marginal Asymptotic Variances and $\hat{l} = 0$ for Electrically Heated Homes in SMUD's Service Area, 1969–1980

Variable	Parameter (1)	Marginal standard error (2)	(1)/(2)
Intercept	-0.2343×10	0.6602	−3.548
Lagged insulation	0.8579	0.5976×10^{-2}	14.356
Real electricity price	0.1818×10^{-2}	0.5955×10^{-3}	3.050
Real income	-0.5744×10^{-1}	0.2458×10^{-1}	−2.337
Usage per HH	0.2989×10^{-1}	0.1178×10^{-1}	2.537
Interest rate	-0.1096	0.2747×10^{-1}	−3.991

(but not necessarily efficient) so long as the matrix of instruments is fixed. See Amemiya, 1974. (Unfortunately, we will find that the nonlinear two-stage least-squares estimator does not exist, in general, when applied to a Box–Cox-transformed system—at least, if we apply it the way it has been proposed in the literature by Amemiya and Powell, 1981.)

C. LIML Results with Marginal Variance for Electrically Heated Homes[26]

The LIML results with marginal variance estimates are shown in Table XXVI. The marginal covariance matrix for the LIML estimates is derived in Appendix C. The $\hat{\sigma}^2$ for the equation is 0.6507×10^{-4} (calculated by dividing by $T - k$), and its marginal variance is of the order of 10^{-15}. Since the computations were carried out with a 12-digit accuracy level, the marginal variance of $\hat{\sigma}^2$ is zero, as is the case with the marginal variance of $\hat{l}$.

The results in Table XXVI are estimated from unscaled data. We estimated also with the data scaled to lie between $0 - 10$ and $0 - 1$. The scaling had to be done prior to transformation in order to allow for subsequent differentiation with respect to l. The estimates except for the intercept are close to those reported in Table XXVI. The picture conveyed by Table XXVI is close to that conveyed by the LIML

[26] This is a technical discussion. The reader may turn to Section VI.D or to the two concluding paragraphs at the end of Section VI.D.

estimates for model 5*, Table XXIV, where the results were reported with a conditional variance. All five variables that enter the model remain statistically significant when the marginal variance, instead of the conditional variance, is used to calculate the ratio of the estimate to the standard error.

In spite of the fact that the results in Table XXVI are bright, we prefer not to use them. We already explained our conceptual difficulties with the marginal variance and what it measures. We add here that our experience with the estimation of the IOLS marginal covariance and our concern with the sample size, round-off errors, and the bounds on the numerical differentiation routine make us weary of relying on estimates of the marginal covariance, as bright as these estimates might be. We use model 5* in Table XXIV instead. We turn now to a few tests of this model.

D. Some Tests of Model 5*, Table XXIV[27]

1. Overidentifying Restrictions

The LIML estimator is known to possess certain desirable properties when the *a priori* restrictions (which include the whole range of assumptions used in estimating the model) are correct. Because of the uncertainty about the *a priori* information, approximate statistical tests have been proposed to test the *a priori* restrictions of the LIML estimation for an overidentified equation. The idea is the following. If all the *a priori* restrictions (taken as a whole) are correct, we would expect the value of the minimum variance ratio used in calculating the LIML estimates not to exceed 1, except as a result of sampling fluctuations. If, on the other hand, the *a priori* restrictions are not correct, we would expect the value of this ratio to exceed 1 by more than what can be accounted for by sampling variability. Hence the appearance of a large value for the minimum variance ratio is cause for rejecting the hypothesis that all the *a priori* assumptions are correct.

But this rejection does not mean that all the assumptions should be rejected at the same time. The test merely sounds a general alarm signaling that "some" *a priori* information used in estimating the model is not correct, but it cannot point to a specific source of the

[27] This section contains parts that may be too technical for some readers. The reader may skip to the concluding paragraph.

trouble. It is up to the investigator to try to track down the source. Unfortunately, the investigator is not free to pursue the search without constraints because the test requires that at least enough restrictions be imposed *a priori* so that the equation to be estimated can be identified. This introduces an element of arbitrariness in the process, since it is necessary to decide which subset of the identifying assumptions will be used as the maintained hypothesis and which subset will be subjected to the test.

We used a test proposed by Basmann (1960). He showed that if the overidentifying restrictions are correct, then the random variable $(\hat{\lambda}_1 - 1)(T - G)/(G - G_1 - m_1)$ will be asymptotically $F_{(G - G_1 - m_1, T - G)}$, where $\hat{\lambda}_1$ is the estimated least-variance ratio. For our model $\hat{\lambda}_1 = 1.251$, $T = 24$, $G = 8$, $G_1 = 5$, $m_1 = 1$. Hence the value of the test statistic is 2.01. This compares with $F_{(2,16)} = 4.45$ at a 5% level of significance. We therefore do not reject the hypothesis that the *a priori* assumptions of our model are correct. But for our model this result is a very approximate one, not only because the test is asymptotic, but because the test presupposes that all of the predetermined variables in the system are exogenous. For this reason, even though our test result does not reject the hypothesis that the *a priori* assumptions of our model are correct, we will still test for two important *a priori* assumptions: normality and the absence of serial correlation.

2. Normality

A plot of the residuals of model 5*, Table XXIV, on a normal probability paper is shown in Fig. 8. (These residuals are the same as the residuals of the model in Table XXVI, since the parameter estimates do not depend on which variance estimate we use.) The plot does not depart significantly from normality. (The W Shapiro–Wilks (1965) statistic does not reject the hypothesis of normality at the 44% signifi-

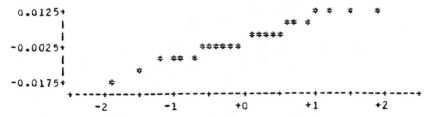

Fig. 8. Normal probability plot of the residuals of model 5*, Table XXIV.

cance level.) This confirms the consistency property of our estimates, which was also implicit in a general way in our test results for the overidentifying restrictions.

3. Serial Correlation

A test of serial correlation in an overidentified structural equation was proposed by Durbin (1957). The test is based on the d statistic, except that in place of the limited information residuals $\hat{u}_{ti}$, the test uses the residuals from the multiple regression of the synthetic dependent variable $\beta_1^{0'} y_t$ on all predetermined variables in the model. Denoting these residuals by z_t^*, the test statistic is

$$d = \frac{\sum_{t=2}^{T} (z_t^* - z_{t-1}^*)^2}{\sum_{t=1}^{T} z_t^{*2}}, \tag{84}$$

which can be checked for significance by using the tables compiled by Durbin and Watson (1951) earlier.

The test results are exact when all the predetermined variables in the system are exogenous, but may be expected to hold approximately for models containing lagged dependent variables. (Godfrey, 1978, proposed an asymptotic test of serial correlation in dynamic simultaneous-equation models, but it requires the equation to be estimated by instrumental variables.)

For our model, $d = 1.57$. Experience with economic time series usually indicates a test against the existence of positive serial correlation. Our calculated d does not indicate the presence of significant positive serial correlation, at the 5% level of significance. (For n = 24 and $k' = 7$, $d_L = 0.75$ at the 5% level of significance. [See Savin and White, 1978.] k' equals the total number of predetermined variables excluding the intercept.) Neither does it indicate the presence of significant negative serial correlation ($4 - 1.57 = 2.43$ is not smaller than d_L and is larger than 2.17 [which is d_U for $k' = 7$]). (See Savin and White, 1977.)

Because of the approximate nature of the test, we also did a visual test of serial correlation. We plotted the limited information residuals $\hat{u}_t$ against $\hat{u}_{t-1}$. The results are shown in Fig. 9. The plot looks random.

We also plotted separately the summer $\hat{u}_t$ against the summer $\hat{u}_{t-1}$. Similarly we plotted the winter $\hat{u}_t$ against the winter $\hat{u}_{t-1}$. In both cases, the scatter diagrams were random. These two plots are not shown here since they add no new information.

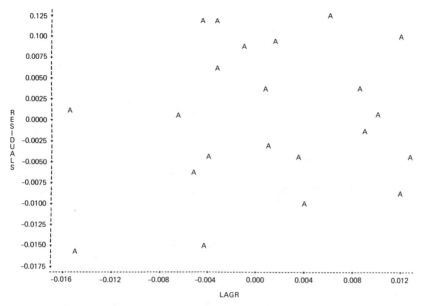

Fig. 9. Plot of the LIML residuals of model 5*, Table XXIV.

Another alternative to the d test is a Lagrangian multiplier (LM) test of the null hypothesis that the serial correlation is zero. This is a much easier test to do than the likelihood ratio test, since it does not entail estimation of the model under the alternative hypothesis and can be shown to be asymptotically equivalent to the likelihood ratio test. We did not do this test partly because our plot showed the residual to be quite random and partly because the LM test results hold only asymptotically anyway.

4. Dynamic Specification

The specification of our model assumes that the adjustment parameter remained unchanged throughout the sample period. On the face of it, this may not be a tenable assumption. Two things could have happened.

First, the coefficient of adjustment may have been subject to a persistent trend during the sample period. In terms of the partial adjustment model, a simple specification that encompasses this possibility is

$$I_{\text{IN}_t} - I_{\text{IN}_{t-1}} = s(I^*_{\text{IN}_t} - (a_0 + a_1 t)I_{\text{IN}_{t-1}}) + u_t, \qquad (85)$$

where t is a trend variable, $t = 1, ..., 24$. Equation (85) leads to

$$I_{IN_t} = sI^*_{IN_t} - b_0 I_{IN_{t-1}} - b_1(t I_{IN_{t-1}}) + u_t, \tag{86}$$

where $b_0 = (1 - sa_0)$, $b_1 = sa_1$, and $I^*_{IN_t}$ is as defined in (2). Equation (86) has an interaction term, t times the lagged dependent variable, which reflects the effect of the trend in the adjustment coefficient. Our specification constrained b_1, the coefficient of this interaction term, to be zero.

Second, the adjustment coefficient may have not been subject to a persistent trend, but it may have been subject to a shift in the post-1973 period, following the jolt of the oil-crisis period. In terms of the partial adjustment model, a simple specification that encompasses this possibility is

$$I_{IN_t} - I_{IN_{t-1}} = s(I^*_{IN_t} - (c_{1t} + c_{2t})I_{IN_{t-1}}) + u_t \qquad (t = 69, ..., 80),$$

$$c_{1t} = \begin{cases} c_1, & 69 \le t \le 73, \\ 0, & \text{otherwise}, \end{cases} \qquad c_{2t} = \begin{cases} c_2, & 74 \le t \le 80 \\ 0, & \text{otherwise} \end{cases} \tag{87}$$

Equation (87) leads to

$$I_{IN_t} = sI^*_{IN_t} - [1 - s(c_{1t} + c_{2t})]I_{IN_{t-1}} + u_t, \tag{88}$$

where $I^*_{IN_t}$ is as defined in (2). Our specification constrained c_{1t} and c_{2t} to be 1. [It should be noted that under the specification in (85) and (87) s is not identifiable.]

To test the specification of the dynamic adjustment, we re-estimated all the equations shown in Table XXIV, but with the model specification changed from (1) to (85) and subsequently from (1) to (87).[28]

Limited information maximum likelihood estimates based on the specification in (85) show that the trend coefficient is not significantly different from zero. The ratio of the estimate of b_1 in (86) to its conditional asymptotic standard error never exceeded 1.7. In particular, the ratio of the trend coefficient for the counterpart of model 5* in Table XXIV to its conditional asymptotic standard error was only 1.1.

What about the possibility of a shift in the coefficient of adjustment

---

[28] Plosser *et al.* (1982) proposed a specification test based on a comparison of the estimates derived from the model and the estimates derived from the differenced form of the same model. To implement the procedure, one needs to estimate by instrumental-variable methods.

in the post-1973 period? The accompanying tabulation lists the LIML estimates of the adjustment coefficients for 1969–1973 and 1974–1980 along with the transformation parameters. These results are based on the specification in (87). The estimates for the two subperiods are so close numerically that we did not feel it is necessary to do a formal test of significance for the differences in the estimated coefficients.

1969–1973	1974–1980	l
0.925	0.927	-1
0.910	0.910	-0.75
0.895	0.893	-0.5
0.883	0.879	-0.25
0.870	0.866	0
0.825	0.828	0.25
0.783	0.811	0.5

5. Insulation Cost

It may be that, contrary to our reasoning and contrary to the follow-up survey results of the California utilities, the insulation cost (other than the interest cost) does affect significantly the household demand for insulation. If that were true, we would normally expect the exclusion of this variable from the model to introduce a systematic element in the behavior of the LIML residuals. In particular, we would expect these residuals to exhibit a positive serial correlation. We found no such evidence, as we have already reported. For a further probe, we plotted the LIML residuals against an index of the real cost of residential construction per square foot. The assumption is that the residential real cost of insulation per square foot is well approximated by the index of real construction per square foot. (The index of construction cost we used was compiled by the Department of Commerce. See Table B.IV, Appendix B. Another index compiled by the Bank of America for Northern California showed almost the same behavior as the index we used. Between 1969 and 1980, Bank of America's index rose by a compounded average rate of 8.7% annually. The Department of Commerce's index rose by 8.8% annually during the same period. These percentages compare to a rise of 7.6% annually in the CPI during the same period.)

Figures 10 and 11 show the results separately for the summer and winter plots of the LIML residuals against the real construction cost index. Both scatter diagrams are random.

We also re-estimated the model with the real interest rate r_t (in 5*, Table XXIV) replaced by

$$\delta C_t + \gamma C_t r_t, \tag{89}$$

where C_t is the real cost of residential construction per square foot. Equation (89) can be rewritten as

$$\delta C_t + \gamma C_t r_t = \gamma C_t(r_t + d), \tag{90}$$

where d is the average annual depreciation rate of insulation. The formulation using (89) was motivated by the fact that when the model is linear, the term $C_t(r_t + d)$ approximates the user's cost. Then the estimates of δ and γ in (89) are related in such a way that

$$\delta = \dot{\gamma} d. \tag{91}$$

Hence an estimate of the depreciation rate can be derived indirectly by making use of (91). This provides a way of estimating the model with the user's cost in the equation when data on the average depreciation rate are not available, as is presently the case. Equation (91) provides us also with a means for checking on the reasonableness of the estimates of δ and γ by examining the reasonableness of the implicit estimate of d. But the simple relationship in (91) does not hold when the estimate of the transformation parameter l is different from 1.

As it turned out, the likelihood function is maximized when $\hat{l} = 0$. The LIML estimates for electrically heated households are listed in the accompanying tabulation.

	Coefficient	Standard deviation condition	t ratio
Intercept	−1.59694	0.51069	−3.1270
Lagged insulation	1.00416	0.09938	10.1039
Real electricity price	0.00175	0.00053	3.3084
Real income/HH	0.00204	0.03942	0.0518
Electricity use/HH	0.00129	0.019601	0.0659
Real cost	−0.02381	0.07155	−0.3328
Real cost × real interest	−0.19052	0.04503	−4.2313

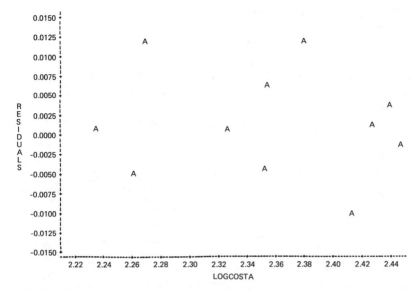

Fig. 10. Plot of the LIML summer residuals against the index of real cost of residential construction per square foot in northern California (1969–1980).

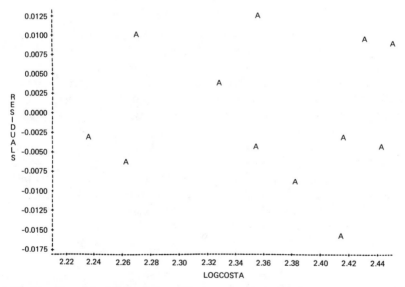

Fig. 11. Plot of the LIML winter residuals against the index of real cost of residential construction per square foot (1969–1980).

The real cost and the real cost × the real interest rate have negative coefficients, but the real cost variable has an insignificant coefficient. (This recurred throughout the various *l* used in transformation.) Together with our preceding results on the cost variable, these results suggest that the statistical significance of the real cost × the real interest is due more to the effect of the real interest rate than the real cost variable.

Note also that, on the whole, the results of estimating with the cost variables are poor: Real income is shown to have no significant effect on the demand for insulation, which is counterintuitive; so is the case with the electricity use per household; lagged insulation has a coefficient larger than 1, implying that *s* in (1) is negative, which makes no economic sense.

In light of these results and in light of the test results reported in this section, we may adopt model 5* of Table XXIV for approximating the demand for insulation by electrically heated households. But before we finalize this step, we turn briefly to one more result. We estimated our model by using nonlinear two-stage (NL2S) least squares as an alternative to LIML. The following section reports the results.

E. Nonlinear Two-Stage Least-Squares Estimation[29]

As we discussed earlier (see Section IV.A) Amemiya and Powell compared the Box–Cox maximum likelihood estimator (BCMLE) with the nonlinear two-stage (NL2S) least-squares estimator and the maximum likelihood estimator (MLE), under the assumption that the untransformed dependent variable follows a two-parameter gamma distribution (Amemiya and Powell, 1981). The BCMLE estimates Amemiya and Powell used for comparison with NL2S are OLS estimates of the model's parameters (with the dependent variable divided by its geometric mean).

The nonlinear two-stage least-squares estimates are derived by minimizing the distance between two projection vectors *y* and *f* (see below) onto the space spanned by X, where X is a matrix of exogenous variables. More formally, consider the model

$$y_t = f(z_{t.}, b.) + u_{t.}, \tag{92}$$

---

[29] The discussion in this section is technical. The reader may turn to the concluding two paragraphs.

where y_t. is a T-component vector, u_t. is a T-component random variable with zero mean and constant variance; z_t. is $T \times H$ matrix consisting partly of endogenous variables and partly of exogenous variables, b. is a G-component vector of unknown parameters, and $f(\cdot)$ is a T-component nonlinear function of, possibly, both z_t. and b.. The NL2S estimator of b. (see Amemiya, 1974) is the value of b. that minimizes

$$h(b) = (y - f)'X(X'X)^{-1}X'(y - f), \qquad (93)$$

where the notations are as defined earlier except that for simplicity we dropped the subscripts from the equation.

When nonlinearities exist only in the parameters, the well-known optimality properties of the two-stage least squares (2SLS) extend to NL2S, under the appropriate choice of the instruments. In particular, the NL2S has the same asymptotic distribution as the LIML and, therefore, attains the smallest asymptotic variance–covariance matrix. But in general these optimality results do not extend to the case in which nonlinearities exist in the variables, as is the case of NL2S estimation of the parameters in the Box–Cox model.

To derive NL2S estimates for the Box–Cox model, Amemiya and Powell minimize a variant of (93), namely,

$$S = (y^* - Xb)'W(W'W)^{-1}W'(y^* - Xb), \qquad (94)$$

with respect to b and l. Here $y^* = (y^l - 1)/l$ is a $T \times 1$ vector of the Box–Cox-transformed dependent variable, X is a $T \times K$ matrix of exogenous variables, and W is a $T \times N$ matrix of instruments such that $X \subset W$. The elements of the remaining columns of W might be low-order polynomials of the columns of X.

Amemiya and Powell did asymptotics for various parameter values in three models; they used BCMLE, NL2S, and MLE under the assumption that the untransformed dependent variable follows a two-parameter gamma distribution. They evaluated the asymptotic mean square error of the BCMLE and the asymptotic variance of NL2S and MLE. The results for the MLE were used only for reference purposes to gauge the best that one could do under the gamma assumption. Their results were not unequivocal in regard to the choice between BCMLE and NL2S. However, the NL2S appeared to have an edge over the BCMLE in the number of cases in which NL2S turned out to be better than BCMLE. The authors recommended that researchers compute the NL2S estimator together with the BCMLE—that is, the estimator proposed initially by Box and Cox. The hope is that this may alert the researcher about large deviations of the BCMLE from its true value.

In our case, the estimates we reported for model 5*, Table XXIV, are LIML estimates and not BCMLE. (Recall, however, that the estimates based on LIML and BCMLE reported for model 6* in Table XI turned out to be close, which suggests that the inconsistency of BCMLE may not be large.) Nonetheless, we will derive NL2S estimates for our model. But before we report estimation results, we should note the following points.

1. In our model, the counterpart for Amemiya and Powell's ($y^* - Xb$) in (94) is the vector of structural-equation residuals. (Amemiya and Powell did not consider the case in which endogenous variables appear in the equation they estimated.) The matrix X in (94) corresponds in our case to the matrix of transformed endogenous and transformed predetermined variables that appear in the equation.

2. The NL2S estimator for the Box–Cox model is easier to work with than the LIML estimator we derived. But the simplicity is due to the simpler model with which Amemiya and Powell worked. In their system, the explanatory variables X were not transformed. Only the dependent variable was transformed. Had the X's been transformed as well, the transformation parameter would have appeared in the W matrix of (94). Since some of the W columns are quadratic (or higher-order) functions of the columns of the (transformed) X's, the derivation of the estimates from (94) will be complicated. It is also not clear that the resulting estimator will be consistent. For this reason, we will let the columns of W be a function of the untransformed exogenous variables (rather than the Box–Cox-transformed version of these variables). This introduces an element of arbitrariness in the NL2S procedure, but that is probably the best we can do at present.

3. The variances of NL2S coefficients estimated by Amemiya and Powell's proposed procedure are all marginal. This means that all the difficulties attending the use (and interpretation) of the marginal estimates discussed earlier will be embodied in these results.

4. But the central problem with the estimator proposed by Amemiya and Powell is that, *in general, there is no finite transformation parameter l that can minimize S in* (94). The criterion will be dominated by the sum of the squares of the residuals $SSR = (y^* -$

$Xb)'(y^* - Xb)$ and will approach zero infinum as $l \to -\infty$ or as $l \to +\infty$, depending on whether all the untransformed dependent variables exceed 1 or are fractions. Let us elaborate.

The difficulty with the S criterion stems from the fact that the scale of the criterion varies with l. For $y_{t.} > 1$, $t = 1, ..., T$, $y^* \to 0$ as $l \to -\infty$; and $SSR \to 0$ as $l \to -\infty$ (see Fig. 12). It is reasonable to expect that for a large enough l, the behavior of SSR will dominate S, since the matrix $M = W(W'W)^{-1}W'$ is constant. Alternatively, it is easy to see that with $y^* \to 0$ as $l \to -\infty$, the choice of $b = 0$ will minimize S. In short, $S \to 0$ as $l \to -\infty$ when $y_{t.} > 1$.

Note that we are not implying that the fit will be better as $l \to -\infty$, but that, in general, the residuals will become *numerically* smaller. Because of the scale effect, S will decline to zero. At the same time the S is declining, the goodness of fit may be improving or may be worsening. What we observe at finite l is the net effect of these two tendencies: the scale effect and the goodness-of-fit effect. Hence, even when S attains a minimum in the interior of a finite interval, it is not clear that the minimum is not spurious. For example, as l decreases (algebraically), the goodness of fit may be deteriorating at a rate faster than the rate of decline of SSR due to

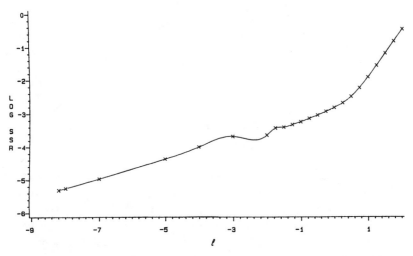

Fig. 12. Log 10 of SSR for NL2SLS estimates for electrically heated homes (SMUD's service area, 1969–1980).

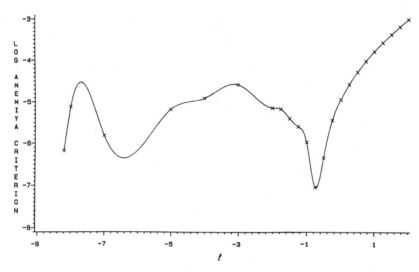

Fig. 13. Log 10 of Amemiya–Powell criterion S for NL2SLS estimates for electrically heated homes (SMUD's service area, 1969–1980).

the scale effect. The net effect of these two tendencies could easily produce a local minimum.

For $y_t. > 1$, we expect $S \to \infty$ as $l \to \infty$. See Fig. 13.

We expect the reverse pattern to prevail when $0 < y_t. < 1$; $SSR \to \infty$ as $l \to -\infty$, and $SSR \to 0$ as $l \to \infty$ (see Fig. 14). Hence S should reveal the same pattern as SSR as l approaches either extreme. See Fig. 15.

Two more observations:

1. When all $y_t.$ are greater than 1 (or when all $y_t.$ are fractions) and when l is numerically small, the shape of S will be affected by the choice of the matrix W. But for large l, the scale effect we just described will dominate.
2. Starting with a given sample $y_t.$, we can divide each element of the sample by an appropriately chosen scaler k such that for one subset of the sample $y_t. > 1$ and for another $0 < y_t. < 1$. Then $S \to \infty$ as $l \to \pm\infty$. In between these extremes, the function should, in general, attain a global minimum. Where this minimum is achieved depends on our choice of k. Since there are infinitely many k's that can scale the sample such that $y_t. > 1$ for one subset

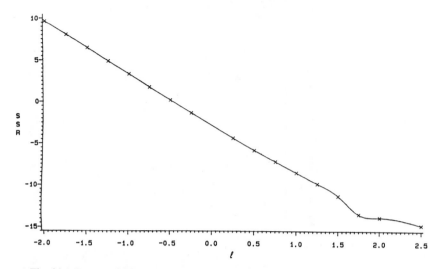

Fig. 14. Log 10 of SSR for NL2SLS estimates with $0 < y_{t_0} < 1$ for electrically heated homes (SMUD's service area, 1969–1980).

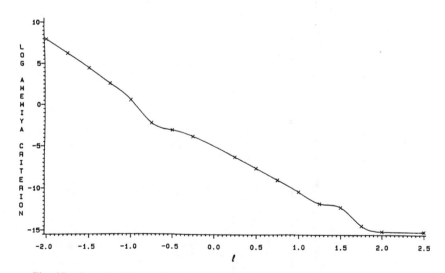

Fig. 15. Log 10 of Amemiya–Powell criterion S for NL2SLS estimates with $0 < y_t. < 1$ for electrically heated homes (SMUD's service area, 1969–1980).

TABLE XXVII

NL2S Conditional Estimates of the Residential Demand for Insulation for All Electrically Heated Homes in SMUD's Service Area, 1969–1980

Equation (1)	Coeff. and t ratio	Intercept (2)	Lagged insulation (3)	Real electricity price, 1974–1980 (4)	Real income per HH (5)	Use per HH (6)	Real interest rate (7)	l (8)	SSR (9)	Value of criterion $S(l)$ (10)
1	Coeff.	−66,906.4	0.927	0.298×10^{-6}	−9.646	133,822.0	-0.662×10^{-4}	−2	0.229×10^{-3}	0.741×10^{-5}
	t	(−1.69)	(13.41)	(1.66)	(−0.41)	(2.58)	(−1.69)			
2	Coeff.	−12,281.6	0.949	0.167×10^{-5}	−4.363	21,497.1	-0.180×10^{-3}	−1.75	0.377×10^{-3}	0.697×10^{-5}
	t	(−1.41)	(13.97)	(2.60)	(−0.40)	(1.41)	(−2.44)			
3	Coeff.	−2,064.8	0.941	0.571×10^{-5}	−2.486	3,099.720	-0.455×10^{-3}	−1.5	0.398×10^{-3}	0.414×10^{-5}
	t	(−1.51)	(14.69)	(2.61)	(−0.58)	(1.51)	(−2.86)			
4	Coeff.	−348.398	0.933	0.204×10^{-4}	−1.405	436.919	-0.116×10^{-2}	−1.25	0.475×10^{-3}	0.266×10^{-5}
	t	(1.495)	(14.51)	(2.73)	(−0.79)	(1.49)	(−3.11)			
5	Coeff.	−60.735	0.924	0.697×10^{-4}	−0.762	61.500	-0.290×10^{-2}	−1	0.578×10^{-3}	0.113×10^{-5}
	t	(−1.48)	(14.21)	(2.81)	(−1.01)	(1.48)	(−3.33)			
6	Coeff.	−11.169	0.913	0.227×10^{-3}	−0.404	8.768	-0.726×10^{-2}	−0.75	0.721×10^{-3}	0.916×10^{-7}
	t	(−1.47)	(13.67)	(2.88)	(−1.24)	(1.48)	(−3.51)			
7	Coeff.	−2.177	0.900	0.685×10^{-3}	−0.211	1.269	-0.180×10^{-1}	−0.5	0.916×10^{-3}	0.473×10^{-6}
	t	(1.47)	(12.92)	(2.94)	(−1.48)	(1.49)	(−3.63)			

290

	Coeff. (t)	Coeff. (t)	Coeff. (t)	Coeff. (t)	Coeff. (t)	Coeff. (t)
8	-0.479 (-1.48)	0.884 (12.02)	0.188×10^{-2} (2.99)	-0.108 (-1.71)	0.187 (1.54)	-0.443×10^{-1} (-3.71)
9	-0.219 (-1.54)	0.866 (10.98)	0.457×10^{-2} (3.05)	-0.546×10^{-1} (-1.90)	0.280×10^{-1} (1.61)	-0.108 (-3.73)
10	-0.338 (-2.01)	0.846 (9.70)	0.981×10^{-2} (3.05)	-0.271×10^{-1} (-2.02)	0.425×10^{-2} (1.69)	-0.261 (-3.64)
11	-0.705 (-2.45)	0.824 (8.03)	0.188×10^{-1} (2.90)	-0.133×10^{-1} (-1.99)	0.647×10^{-3} (1.68)	-0.623 (-3.34)
12	-1.450 (-2.38)	0.800 (6.08)	0.331×10^{-1} (2.54)	-0.644×10^{-2} (-1.78)	0.980×10^{-4} (1.55)	-1.474 (-2.80)
13	-2.916 (-1.99)	0.772 (4.26)	0.550×10^{-1} (2.07)	-0.312×10^{-2} (-1.48)	0.147×10^{-4} (1.32)	-3.451 (-2.18)
14	-5.800 (-1.54)	0.740 (2.85)	0.886×10^{-1} (1.61)	-0.151×10^{-2} (-1.18)	0.219×10^{-5} (1.08)	-7.980 (-1.62)
15	-11.440 (-1.14)	0.703 (1.87)	0.140 (1.23)	-0.728×10^{-3} (-0.92)	0.323×10^{-6} (0.87)	-18.150 (-1.18)
16	-22.290 (-0.83)	0.660 (1.21)	0.220 (0.93)	-0.351×10^{-3} (-0.73)	0.473×10^{-7} (0.70)	-40.247 (-0.85)
17	-42.468 (-0.58)	0.611 (0.78)	0.345 (0.71)	-0.169×10^{-3} (-0.57)	0.687×10^{-8} (0.56)	-86.749 (-0.59)

-0.25	0.118×10^{-2}	0.372×10^{-5}
0	0.155×10^{-2}	0.118×10^{-4}
0.25	0.214×10^{-2}	0.273×10^{-4}
0.5	0.331×10^{-2}	0.540×10^{-4}
0.75	0.603×10^{-2}	0.977×10^{-4}
1	0.127×10^{-1}	0.167×10^{-3}
1.25	0.289×10^{-1}	0.273×10^{-3}
1.5	0.672×10^{-1}	0.432×10^{-3}
1.75	0.156	0.665×10^{-3}
2.0	0.356	0.977×10^{-3}

and $0 < y_{t.} < 1$ for another, there will be infinitely many minima, each depending on our (arbitrary) choice of k. We note that a similar scaling problem does not arise in the search for the LIML estimates, because the scale of the likelihood function is invariant with respect to any normalization of the variables.

It is not within the scope of this study to develop an appropriately modified NL2S to overcome the scaling problem posed by the NL2S (see the discussion in Khazzoom, 1984, pp. 156 ff). We report our NL2S results for illustrative purposes only.

Except for its last two columns, Table XXVII shows the results of the NL2S conditional estimates in essentially the same format as we used earlier in reporting estimation results. In the last two columns we report the sum of the squares of the residuals (SSR) and the value of the criterion to be minimized (S). We used six instruments for estimation: the three vectors of exogenous variables that appear in the equation and the three vectors derived by squaring the elements of these three exogenous variables. Except for the intercept, the NL2S estimates are, roughly speaking, not very different from the LIML's estimates when $l \in (-1, 0.5)$, although substantial differences in the estimates of the price parameter begin to appear already when l takes 0 and larger values. In general, the LIML price coefficients tend to be more stable than the NL2Ss.

Figures 12 and 13 plot the results shown in columns 8 and 9 over a wider range of l. To smooth the plot, the routine uses a cubic spline. It can be been that as l decreases from 2 to -8, SSR decreases to zero (Fig. 12); S decreases as l decreases, attaining a minimum around $l = -0.75$ (Fig. 13). The function rises and then declines as l continues to decline and attains a second local minimum around $l = -6.5$. It rises beyond that and then declines again. We tried to evaluate the criterion at $l = -12$, but the program failed to carry out the estimation of the parameters. In fact, the lowest value of l for which we were able to get the program to calculate the estimates was -8.2.

Figures 14 and 15 plot SSR and S for the case in which the untransformed dependent variable is a fraction. To convert $y_{t.}$ to positive fractions we divided $y_{t.}$ (and all the explanatory variables) by the geometric average of all endogenous variables in the system. This is the same geometric average by which all variables had to be divided when we used LIML estimation in order to allow for the Jacobian. [The estimates provide a concrete illustration of the fact that, unlike the

LIML estimates which (except for the intercept) are invariant when the untransformed variables are divided by a constant, the NL2S estimates do not possess this property.]

Figure 14 shows that SSR declines to 0 as l increases to infinity. (This is the opposite pattern we noticed in Table 12, where $y_{t.} > 1$.) As l increases, S drifts to zero (Fig. 15). We attempted to extend the estimation to l's beyond those shown in Fig. 15. But for negative and numerically large l, the program could not calculate results, since the SSRs were extremely large. For positive but numerically large l, the program could not calculate results because the SSRs were too small and were apparently rounded off to zero.

To conclude, the NL2S estimates illustrate the type of results we could get if we were to use NL2S instead of the LIML procedure. The basic difficulty is that the NL2S proposed by Amemiya and Powell for a Box–Cox model does not exist in general. Conceivably, the problem can be overcome, and the estimator can be salvaged by an appropriate standardization of the criterion (since the behavior of the criterion is due to the scaling problem introduced by the Box–Cox transformation). But the determination of the optimal standardization procedure, the analytical behavior of the standardized criterion, and the properties of the resulting estimator need to be investigated before the NL2S or a standardized version of it can be used in estimation.

For these practical considerations, and in light of the good performance of the LIML estimates of our model when subjected to tests, we adopt model 5* shown in Table XXIV as our model for approximating the household demand for home insulation.

VII. SUMMARY

A. Overview

1. There has been a growing need over the past few years for the development of a behavioral model of the demand for conservation. This happened as the utilities, as a group, have become increasingly a supplier of conservation in the same sense that, as a group, they are a supplier of energy. Consequently, there emerged a need to plan for the demand for conservation in the same way that there was a need to plan for the demand for energy:

   But the interest in behavioral modeling of the demand for

conservation was also heightened by the increased realization that in the search for an optimal investment strategy, the utility can improve the efficiency of allocation of its investment resources by opting for an active role in shaping the demand it faces for both energy and conservation, rather than playing the passive role of merely responding to changes in outside conditions. The development of a system which estimates behaviorally the demand for conservation provides the utility (and the regulatory agency) with a tool that enables it to estimate the household response to measures intended to affect the household demand for conservation, one way or another. These measures may originate with the utility (changes in the price of energy, subsidy of the interest rate, rebates, etc.) or the state and federal government (e.g., tax credits).

This study is a first step toward meeting this need and providing us with the capability of estimating behaviorally the household demand for one major component of the demand for conservation, namely, the demand for home insulation.

2. The demand relationships we estimate provide one block (out of three) in a jointly determined system of demand relationships: demand for electricity, demand for insulation, and demand for efficient appliances. Figure 16 provides an overview of the flow of influence among the three blocks.

3. The study is pitched toward the service-area level. We develop and estimate a behavioral model of the household demand for insulation in the Sacramento Municipal Utility District's (SMUD's) service area. The method we use to generate the time series of the insulation index as well as the econometric methods we use to estimate this model can be adapted easily to other service areas as well. They can also be extended without much difficulty to geographic areas larger than just one service area.

4. There is no generally agreed-upon solution for dealing with the uncertainty in the choice of the "appropriate" design matrix. Under the present state of the art, the solution must rest on pragmatic grounds. The approach in this study is based on identifying, at the outset, a suitably parsimonious model by drawing as much as possible on all available *a priori* information—theoretical as well as institutional—to narrow down the range of admissible variables.

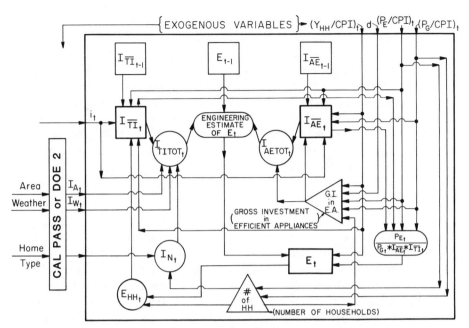

Fig. 16. Interrelationship among the demand for thermal integrity, the demand for efficient appliances, and the demand for electricity. [Abbreviations: d, depreciation of appliances; E, electricity usage; E_{HH}, electricity usage per household; I_A, index of dwelling area (square footage); I_{AE}, index of appliance efficiency; I_{AETOT}, index of total effect of appliance efficiency (same as I_{AE} but taking into account, in addition, the stock and the number of households); I_{TI}, index of thermal integrity; I_{TITOT}, index of total effect of thermal integrity (same as I_{TI} but taking into account, in addition, the square footage, the number of homes, and wealth); I_{N_t}, index of number of dwellings; P_E, price of electricity; P_G, price of gas; I_W, index of weather conditions; Y_{HH}, income per household.

5. We use first an engineering model (CAL PASS) to provide the data base necessary to generate the time series for the insulation index. We calculate the index for 1960–1980 and for each type of dwelling (single-family, duplex, multiple-family, and mobile homes) with each dwelling classified by heating and cooling mode (resistance heat, heat pump, gas heat, heat-pump cooling, air conditioning). The model can be estimated at any desired level of aggregation, so long as the required energy sales data are available in a comparable breakdown. An aggregate insulation index that meets desirable properties can be derived for any level of aggregation required.

The method we use to generate the index of insulation demand can be applied in estimating the demand for any conservation activity that affects the thermal integrity of the home. Examples include double glazing, installation of hangover, and changing the direction of windows away from or toward the sun.

6. Our sample period is 1969–1980. We estimate first for two categories—households with electrically heated homes and households with gas-heated homes. But our focus is on electrically heated homes. Our estimates separate also the summer and winter demand for insulation.

B. Method

7. We use a Box–Cox transform for both dependent and explanatory variables and estimate our model first as a one-equation model using iterative ordinary least squares (IOLS) and bootstrap methods. We then extend the limited information maximum likelihood (LIML) method to a Box–Cox-transformed system to derive consistent estimates for our model. We also estimate by using a nonlinear two-stage least-squares estimator proposed in the literature for estimating a Box–Cox model. In the process, we address methodological issues raised by the use of Box–Cox-transformed system.

C. IOLS Results

8. The iterative ordinary least-squares (IOLS) results for the winter months for electrically heated dwellings show that in the post-oil-crisis years the price of electricity had a positive and a statistically significant effect on the demand for insulation. The estimates show also that the level of electricity usage by household had a positive effect on the demand for insulation. This is as it should be, since the greater the electricity usage, the greater the benefit from insulation. The estimates also show that insulation is an inferior good (that is, a good whose demand decreases as income rises) and that higher interest rates have had an increasingly negative and statistically significant effect on the demand for insulation.

9. Our results also show that the demand for insulation tends to lag behind changes in the economic conditions. For electrically heated homes, the median lag is approximately five years.

10. Of all factors affecting the demand for insulation, the interest rate has the greatest impact. This should not come as a surprise, since the demand for insulation is a demand for capital. The average long-run interest rate elasticity of insulation demand is close to 1. This is eight times as large as the long-run elasticity of insulation demand with respect to the price of electricity.

    These results provide a basis for the argument that many have made for subsidizing the interest rate on loans to finance insulation. (Strictly, however, these results, as supportive as they turned out to be, provide only the necessary but not sufficient conditions for a policy of interest rate subsidy.) Our results also appear to provide support for the argument of some regulatory agencies that the effect of energy price on the demand for insulation may be too small in magnitude to be solely relied upon to increase the demand for home insulation.

    The summer results for the electrically heated homes are essentially the same as the winter results, except for the fact that, not surprisingly, the summer demand elasticity with respect to electricity price tends to be smaller than the corresponding winter price elasticity.

11. For gas-heated homes, we derived IOLS estimates for the summer months only. These results are qualitatively the same as those derived for electrically heated homes. But they (the results for gas-heated homes) should be taken with a great deal of caution because of the small sample size for gas-heated homes.

D. Implications for IOLS Results

12. The results reported in points 8–11 suggest that so long as the interest rate continues to decline and so long as the real price of electricity continues to increase, we are likely to witness a continuation of the recent upward trend we have witnessed in the household's demand for insulation. In contrast, the increases in real income will exert a downward pressure on the demand for insulation. Similarly, the reduction in the real price of gas forecast to take place in the future will have the same downward

effect on insulation demand. But the retardation due to the drop in gas prices is not likely to be strong because of the relatively small elasticity of insulation demand with respect to the price of gas.

13. Overall, these results suggest that households do respond on their own to economic stimuli. In effect, our results put in question the argument many have made that without government mandating of standards of thermal integrity and, in general, without government intervention, there will be hardly any conservation by the public. Our results were calculated for a period in which there was hardly any government involvement in conservation. They demonstrate that households know their calculus and that they do respond substantially to changes in the economic conditions, as economic theory would lead one to believe. This is significant, since there has been a great deal of concern recently over what would happen to the conservation of energy if government were to pull out of its role of mandating conservation standards. The answer our results suggest is that nothing would happen: As long as the economic conditions continue to favor conservation, households will continue their conservation efforts, at least those efforts pitched toward reducing space heating and air conditioning costs.

E. The Marginal-Variance and Bootstrap Estimation

14. In connection with models that use a Box–Cox transformation, it has been argued in the literature that the conventional method of calculating the variance of the estimate is not appropriate and that it tends to exaggerate the stability of the estimated parameters. An alternative procedure was proposed in the literature, but there is controversy about the scientific relevance of this alternative. (This has come to be known as the conditional-versus-marginal-variance controversy.) We note in passing that the controversy is not unique to Box–Cox-transformed systems. It permeates several other statistical endeavors.

In this study, we report results based on both methods of calculating the variance, even though we view conditional variance as more appropriate. In addition, we attempt to achieve further insights into our results by using bootstrap estimation methods.

F. Consistent Estimation of the Model

15. Because of the fact that the demand for insulation is jointly determined with the demand for electricity, we went beyond the IOLS and estimated our model by using consistent methods (these are methods that are suitable for jointly determined systems). In particular, we extended the limited information maximum likelihood (which is a consistent method) to apply to a Box–Cox-transformed system. We also subjected our estimates to tests of the overidentifying restrictions, normality, serial correlation, and specification.

16. In light of its good performance when subjected to these various tests, we adopted model 5* of Table XXIV for estimating the household demand for home insulation.

APPENDIX A: DERIVATION OF THE INFORMATION MATRIX FOR THE IOLS

Given the log likelihood

$$L^* = -\frac{T}{2}\log(2\pi) - T\log\sigma - (2\sigma^2)^{-1}(Y^{(l)} - X^{(l)}a)'(Y^{(l)} - X^{(l)}a)$$

$$+ (l-1)\sum_{t=1}^{T}\log Y_t, \tag{A.1}$$

define

$$S = \sum_{t=1}^{T}\left(Y_t^{(l)} - \sum_{j=1}^{p} X_{tj}^{(l)} a_j\right)^2, \tag{A.2}$$

$$\frac{\partial L^*}{\partial a} = -(2\sigma^2)^{-1}\left(\frac{\partial S}{\partial a_k}\right) = -(2\sigma^2)^{-1}$$

$$\times \left[2\sum_{t=1}^{T}\left(Y_t^{(l)} - \sum_{j=1}^{p} X_{tj}^{(l)} a_j\right)(-X_{tk}^{(l)})\right]$$

$$= \frac{-X^{(l)\prime}X^{(l)}a + X^{(l)\prime}Y^{(l)}}{\sigma^2}, \tag{A.3}$$

$$E\left(-\frac{\partial^2 L^*}{\partial a^2}\right) = \frac{X^{(l)\prime}X^{(l)}}{\sigma^2}, \tag{A.4}$$

$$E\left(-\frac{\partial^2 L^*}{\partial\sigma\,\partial a}\right) = E[-2\sigma^{-3}(X^{(l)\prime}X^{(l)}a - X^{(l)\prime}Y^{(l)})]$$

$$= 0 \quad \text{by (A.3),} \tag{A.5}$$

$$E\left(-\frac{\partial^2 L^*}{\partial l\,\partial a_k}\right) = \sigma^{-2}E\left[\sum_{t=1}^{T}\left(Y_t^{(l)} - \sum_{j=1}^{p}X_{tj}^{(l)}a_j\right)\left(\frac{-\partial X_{tk}^{(l)}}{\partial l}\right)\right.$$

$$\left. + \sum_{t=1}^{T}\left(\frac{\partial Y_t^{(l)}}{\partial l} - \sum_{j=1}^{p}\frac{\partial X_{tj}^{(l)}}{\partial l}a_j\right)(-X_{tk}^{(l)})\right]$$

$$= 0 + \sigma^{-2}\sum_{t=1}^{T}E\left[-\frac{Y_t^{(l)}}{l} + \frac{Y_t^l\ln Y_t^l}{l^2} + \sum_{j=1}^{p}\frac{X_{tj}^{(l)}a_j}{l}\right.$$

$$\left. - \sum_{j=1}^{p}\frac{X_{tj}^l\ln X_{tj}^l a_j}{l^2}\right]\cdot[-X_{tk}^{(l)}]$$

$$= \sigma^{-2}\sum_{t=1}^{T}\left[E\left(\frac{Y_t^l\ln Y_t^l}{l^2}\right) - \sum_{j=1}^{p}\frac{X_{tj}^l\ln X_{tj}^l a_j}{l^2}\right]\cdot[-X_{tk}^{(l)}]$$

$$= \sigma^{-2}\sum_{t=1}^{T}\left[E\left(-\frac{V_t\ln V_t}{l^2} + \sum_{j=1}^{p}\frac{X_{tj}^l\ln X_{tj}^l a_j}{l^2}\right)\right.$$

$$\left. \cdot[-X_{tk}^{(l)}] \right. \tag{A.6}$$

In matrix form:

$$E\left(\frac{-\partial^2 L^*}{\partial l\,\partial a}\right) = \sigma^2\begin{bmatrix} X_{11}^{(l)} & X_{12}^{(l)} & \cdots & X_{1p}^{(l)} \\ \vdots & \vdots & & \vdots \\ X_{T1}^{(l)} & X_{T2}^{(l)} & \cdots & X_{Tp}^{(l)} \end{bmatrix}$$

$$\cdot\left\{\begin{bmatrix} \dfrac{-E(V_1\ln V_1)}{l^2} \\ \vdots \\ \dfrac{-E(V_T\ln V_T)}{l^2} \end{bmatrix} + \begin{bmatrix} \displaystyle\sum_{j=1}^{p}\dfrac{X_{1j}^l\ln X_{1j}^l a_j}{l^2} \\ \vdots \\ \displaystyle\sum_{j=1}^{p}\dfrac{X_{Tj}^l\ln X_{Tj}^l a_j}{l^2} \end{bmatrix}\right\}$$

$$= \sigma^{-2}X^{(l)\prime}b, \tag{A.6'}$$

where b is the $T \times 1$ vector

$$b = \frac{-E(V_t \ln V_t)}{l^2} + \sum_{j=1}^{p} \frac{X'_{tj} \ln X'_{tj} a_j}{l^2} \qquad (t = 1, \ldots, T)$$

and

$$\frac{\partial Y_t^{(l)}}{\partial l} = \frac{Y^l \ln Y_t^l - (Y^l - 1)}{l^2}, \qquad \frac{\partial X_{tj}^{(l)}}{\partial l} = \frac{X'_{tj} \ln X'_{tj} - (X'_{tj} - 1)}{l^2}$$

and

$$V_t = Y_t^l \sim N\left(1 + l \sum_{j=1}^{p} X_{tj}^{(l)} a_j, \; l^2 \sigma^2\right), \qquad (A.7)$$

$$\frac{\partial L^*}{\partial l} = -(2\sigma^2)^{-1} \frac{\partial S}{\partial l} + \sum_{t=1}^{T} \ln Y_{t.}$$

$$= -\sigma^{-2} \sum_{t=1}^{T} \left(Y_t^{(l)} - \sum_{j=1}^{p} X_{tj}^{(l)} a_j\right)$$

$$\times \left(\frac{\partial Y_t^{(l)}}{\partial l} - \sum_{j=1}^{p} \frac{\partial X_{tj}^{(l)}}{\partial l} a_j\right) + \sum_{t=1}^{T} \ln Y_t = 0, \qquad (A.8)$$

$$E\left(-\frac{\partial^2 L^*}{\partial \sigma \, \partial l}\right) = -\sigma^{-3} E\left(\frac{\partial S}{\partial l}\right) = -\sigma^{-3} E(2\sigma^2) \sum_{t=1}^{T} \ln Y_t$$

$$= \frac{-2}{l\sigma} E \sum_{t=1}^{T} \log|V_t|, \qquad (A.9)$$

$$E\left(-\frac{\partial^2 L^*}{\partial l^2}\right) = (2\sigma^2)^{-1} \frac{\partial^2 S}{\partial l^2}$$

$$= \sigma^{-2} E\left[\sum_{t=1}^{T} \left(\frac{\partial Y_t^{(l)}}{\partial l} - \sum_{j=1}^{p} \frac{\partial X_{tj}^{(l)}}{\partial l} a_j\right)^2\right.$$

$$\left. + \sum_{t=1}^{T} \left(Y_t^{(l)} - \sum_{j=1}^{p} X_{tj}^{(l)} a_j\right)\left(\frac{\partial^2 Y_t^{(l)}}{\partial l^2} - \sum_{j=1}^{p} \frac{\partial^2 X_{tj}^{(l)}}{\partial l^2} a_j\right)\right]$$

$$= \sigma^{-2} C, \qquad (A.10)$$

where

$$C = \sum_{t=1}^{T} \left[E \left(\frac{V_t \ln|V_t| - (V_t - 1)}{l^2} - \sum_{j=1}^{p} \frac{X_{tj}^l \ln|X_{tj}^l| - (X_{tj}^l - 1)}{l^2} a_j \right)^2 \right.$$

$$+ E \left(\frac{V_t - 1}{l} - \sum_{j=1}^{p} \left(\frac{X_{tj}^l - 1}{l} \right) a_j \right)$$

$$\left. \times \left(\frac{lV_t(\ln V_t)^2 - 2l(V_t \ln|V_t| - V_t + 1)}{l^4} \right) \right],$$

and

$$\frac{\partial^2 Y_t^{(l)}}{\partial l^2} = \frac{l^3 Y_t^l(\ln Y_t)^2 - 2l(lY_t^l \ln Y_t - Y_t^l + 1)}{l^4}, \tag{A.11}$$

$$\frac{\partial L^*}{\partial \sigma} = -T\sigma^{-1} + \sigma^{-3}(Y^{(l)} - X^{(l)}a)'(Y^{(l)} - X^{(l)}a) = 0, \tag{A.12}$$

$$E - \left(\frac{\partial^2 L}{\partial \sigma \, \partial \sigma} \right) = E[-T\sigma^{-2} + 3\sigma^{-4}E(Y^{(l)} - X^{(l)}a)'(Y^{(l)} - X^{(l)}a)]$$

$$= -T\sigma^{-2} + 3\sigma^{-4}(T\sigma^2) = 2T\sigma^{-2}. \tag{A.13}$$

APPENDIX B: LISTING OF ADDITIONAL TABLES USED IN ESTIMATION

TABLE B.I

Tail-End Block of Electricity Price (Cents per Kilowatt Hour) and Gas Price (Dollars per Therm) in SMUD's Service Area[a]

| | Electrically heated homes | | | | Gas-heated homes | | | |
| | Winter | | Summer | | Winter | | Summer | |
Year	Elec-tricity	Gas	Elec-tricity	Gas	Elec-tricity	Gas	Elec-tricity	Gas
1968	0.50	0.0611	0.90	0.0611	0.70	0.0611	0.90	0.0611
1969	0.50	0.0625	0.90	0.0625	0.70	0.0625	0.90	0.0625
1970	0.50	0.0703	0.90	0.0703	0.70	0.0703	0.90	0.0703
1971	0.50	0.0712	1.00	0.0712	0.80	0.0712	1.00	0.0712
1972	0.50	0.0752	1.12	0.0752	0.88	0.0752	1.12	0.0752
1973	0.60	0.0831	1.12	0.0831	0.88	0.0831	1.12	0.0831
1974	0.60	0.0939	1.24	0.0989	1.24	0.0989	1.24	0.0989
1975	0.67	0.1374	1.24	0.1374	1.24	0.1374	1.24	0.1374
1976	0.67	0.1588	1.60	0.1588	1.22	0.1588	1.60	0.1588
1977	0.86	0.2231	1.60	0.2231	1.22	0.2231	1.60	0.2231
1978	0.86	0.3134	1.96	0.3134	1.60	0.3134	1.96	0.3134
1979	1.12	0.4404	1.96	0.4404	1.60	0.4404	1.96	0.4404
1980	1.12	0.5330	2.60	0.5330	2.20	0.5330	2.60	0.5330
1981	1.71	0.6088	2.91	0.6088	2.47	0.6088	2.91	0.6088

[a] Source: SMUD's files.

TABLE B.II

Economic Data[a]

Year	CPI	AAA	CPIAP	YTOTR	YPC : W = YTOTR ÷ total number of winter customers	YPC : S = YTOTR ÷ total number of summer customers
1968	104.2	6.175	102.10	2,096,524,105	11,391.0	11,391.0
1969	109.8	7.029	104.30	2,119,138,512	11,218.0	11,260.9
1970	116.3	8.040	106.90	2,189,414,543	11,215.9	11,208.7
1971	121.3	7.387	109.10	2,299,965,535	11,310.2	11,316.9
1972	125.3	7.213	109.70	2,442,363,634	11,453.6	11,409.3
1973	131.1	7.441	109.80	2,567,546,776	11,533.8	11,534.2
1974	147.1	8.566	115.60	2,560,909,773	11,090.2	11,089.7
1975	161.2	8.830	128.10	2,599,995,624	10,868.1	10,859.8
1976	170.5	8.430	135.30	2,722,108,694	10,994.2	10,987.6
1977	181.5	8.020	140.10	2,958,126,186	11,449.0	11,422.7
1978	195.4	8.730	150.10	3,130,682,931	11,563.9	11,530.3
1979	217.4	9.630	154.93	3,209,944,663	11,329.2	11,297.7
1980	246.8	11.940	163.12	3,068,342,003	10,329.9	10,392.7
1981	272.4	14.170	175.59	3,269,409,347	10,783.0	10,764.6

Abbreviations and Sources: CPI: Consumer Price Index for all urban consumers, U.S. city average, all items; source: Council of Economic Advisors, *Economic Report of the President* (Washington, D.C.: CEA, Executive Office of the President, February, 1982). Table B-52. AAA: Moody's AAA bond rate; source: Bureau of Economic Analysis, *Survey of Current Business* (Washington, D.C. Department of Commerce), monthly issues. CPIAP: Consumer Price Index of Appliances, for all urban consumers, U.S. city average; source: Bureau of Labor Statistics, *Monthly Labor Review* (Washington, D.C.: Department of Labor), monthly issues. YTOTR: total real personal income, Sacramento SMSA; source: Bureau of Economic Analysis, Regional Economics Information System (Washington, D.C.: Department of Commerce), unpublished data, current as of April 1982.

TABLE B.III

Annual Gas Consumption, in Therms per Household in Sacramento County[a]

Year	Sacramento City	Sacramento County
1968	1057	1112
1969	1120	1180
1970	1030	1078
1971	1173	1224
1972	1113	1165
1973	1062	1116
1974	997	1035
1975	1078	1122
1976	925	962
1977	845	869
1978	816	839
1979	832	861
1980	768	783
1981	667	684

[a] Source: PG&E's files.

TABLE B.IV

Index of Residential Construction Cost per Square Foot in San Francisco (1913 = 100%)[a]

Year	Index	Year	Index
1968	966	1975	1698
1969	1054	1976	1906
1970	1088	1977	2063
1971	1174	1978	2263
1972	1285	1979	2498
1973	1469	1980	2671
1974	1552	1981	2873

[a] Source: 1968–1978: Department of Commerce, Bureau of Economic Analysis, "Construction and Real Estate—Housing Starts and Permits, Construction Cost Indexes," *Business Statistics,* 1979 Edition, p. 45. 1979: Department of Commerce, *Survey of Current Business* (February, 1981) Vol. 61(2), page S.9. 1980–1981: Department of Commerce, *Survey of Current Business* (December, 82) Vol. 62(12), page S-7.

APPENDIX C

Derivation of the Information Matrix for the LIML
Estimates of a Box–Cox-Transformed System

Refer to Eq. (81) and focus on the first equation in (75). In the following we derive the elements of the matrix in (82). We have

$$\frac{\partial L^*}{\partial \gamma_{.1}} = -\frac{T}{2}\frac{1}{\sigma_{11}}(-2M_{31}^{(l)}\beta_{.1}^0 + 2M_{33}^{(l)}\gamma_{.1}), \qquad (C.1)$$

$$\frac{\partial^2 L^*}{\partial \gamma_{.1}\,\partial \beta_{.1}^{0'}} = \frac{T}{\sigma_{11}}\,M_{31}^{(l)}. \qquad (C.2)$$

From (82), we have

$$S_{\beta_1^0 \gamma_{.1}} = -\plim_{T\to\infty} M_{13}^{(l)} \qquad (C.3)$$

$$\frac{\partial^2 L^*}{\partial \gamma_{.1}\,\partial \gamma_{.1}'} = -\frac{T}{\sigma_{11}}\,M_{33}^{(l)}, \qquad (C.4)$$

$$\therefore\; S_{\gamma_1 \gamma_{.1}} = \plim_{T\to\infty} M_{33}^{(l)}; $$

$$\frac{\partial^2 L^*}{\partial \gamma_{.1}\,\partial \sigma_{11}} = \frac{T}{2}\frac{1}{\sigma_{11}^2}(-2M_{31}^{(l)}\beta_{.1}^0 + 2M_{33}^{(l)}\gamma_{.1}), \qquad (C.6)$$

$$\therefore\; S_{\gamma_{.1}\sigma_{11}} = 0 \quad \text{by} \;\;(C.1); \qquad (C.7)$$

$$\frac{\partial^2 L^*}{\partial \gamma_{.1}\,\partial l} = \frac{T}{\sigma_{11}}(D_{31}^{(l)}\beta_{.1}^0 - D_{33}^{(l)}\gamma_{.1}), \qquad (C.8)$$

where

$$D_{ij}^{(l)} = \partial M_{ij}^{(l)}/\partial l. \qquad (C.9)$$

$D_{31}^{(l)}$ is the $G_1 \times (m_1 + 1)$ matrix with the rsth term given by

$$\frac{1}{T}\sum_{t=1}^{T}\left(x_{tr}^{(l)}\frac{\partial y_{ts}^{(l)}}{\partial l} + y_{ts}^{(l)}\frac{\partial x_{tr}^{(l)}}{\partial l}\right).$$

$D_{33}^{(l)}$ is the $G_1 \times G_1$ matrix with the rsth term given by

$$\frac{1}{T}\sum_{t=1}^{T}\left(x_{tr}^{(l)}\frac{\partial x_{ts}^{(l)}}{\partial l} + x_{ts}^{(l)}\frac{\partial x_{tr}^{(l)}}{\partial l}\right).$$

x_{tr} is the tth observation on the rth predetermined variable in the equation, $r = 1, 2, ..., G_1$. Y_{ts} is the tth observation on the sth endogenous variable in the equation, $s = 1, 2, ..., m_1 + 1$.

Note that

$$\frac{\partial y_{ts}^{(l)}}{\partial l} = \frac{ly_{ts}^l \ln y_{ts} - y_{ts}^l + 1}{l^2};$$

$\partial x_{tr}^{(l)}/\partial l$ has a similar expression (replace y_{ts} by x_{tr} above).

From (C.8), it follows that

$$S_{\gamma_{.1}l} = \underset{T \to \infty}{\text{plim}}(D_{33}^{(l)}\gamma_{.1} - D_{31}^{(l)}\beta_{.1}^0), \tag{C.10}$$

$$\frac{\partial L^*}{\partial \beta_{.1}^0} = \frac{T}{2} \cdot \frac{2W_{11}^{(l)}\beta_{.1}^0}{\beta_{.1}^{0\prime}W_{11}^{(l)}\beta_{.1}^0} - \frac{T}{2} \cdot \frac{1}{\sigma_{11}}(2M_{11}^{(l)}\beta_{.1}^0 - 2M_{13}^{(l)}\gamma_{.1}), \tag{C.11}$$

$$\frac{\partial^2 L^*}{\partial \beta_{.1}^0 \, \partial \beta_{.1}^{0\prime}} = T \cdot \frac{(\beta_{.1}^{0\prime}W_{11}^{(l)}\beta_{.1}^0)W_{11}^{(l)} - 2W_{11}^{(l)}\beta_{.1}^0\beta_{.1}^{0\prime}W_{11}^{(l)}}{(\beta_{.1}^{0\prime}W_{11}^{(l)}\beta_{.1}^0)^2}$$

$$- \frac{T}{\sigma_{11}}M_{11}^{(l)}, \tag{C.12}$$

$$S_{\beta_{.1}^0\beta_{.1}^0} = \underset{T \to \infty}{\text{plim}}\left[M_{11}^{(l)} - \sigma_{11}\frac{W_{11}^{(l)}}{\beta_{.1}^{0\prime}W_{11}^{(l)}\beta_{.1}^0} + 2\sigma_{11}\frac{W_{11}^{(l)}\beta_{.1}^0\beta_{.1}^{0\prime}W_{11}^{(l)}}{(\beta_{.1}^{0\prime}W_{11}^{(l)}\beta_{.1}^0)^2}\right]$$

$$= \left(\begin{array}{c} \cdot \\ S_{\beta_{.1}\beta_{.1}} \end{array}\right), \tag{C.13}$$

$$\frac{\partial^2 L}{\partial \beta_{.1}^0 \, \partial \sigma_{11}} = \frac{T}{\sigma_{11}^2}(M_{11}^{(l)}\beta_{.1}^0 - M_{13}^{(l)}\gamma_{.1}), \tag{C.14}$$

$$S_{\beta_{.1}^0\sigma_{11}} = -\frac{1}{\sigma_{11}}\underset{T \to \infty}{\text{plim}}(M_{11}^{(l)}\beta_{.1}^0 - M_{13}^{(l)}\gamma_{.1}) = \left(\begin{array}{c} \cdot \\ S_{\beta_{.1}\sigma_{11}} \end{array}\right), \tag{C.15}$$

$$\frac{\partial^2 L^*}{\partial \beta_{.1}^0 \, \partial l} = T \cdot \frac{(\beta_{.1}^{0\prime}W_{11}^{(l)}\beta_{.1}^0)\frac{\partial W_{11}^{(l)}}{\partial l}\beta_{.1}^0 - \left(\beta_{.1}^{0\prime}\frac{\partial W_{11}^{(l)}}{\partial l}\beta_{.1}^0\right)W_{11}^{(l)}\beta_{.1}^0}{(\beta_{.1}^{0\prime}W_{11}^{(l)}\beta_{.1}^0)^2}$$

$$- \frac{T}{\sigma_{11}}(D_{11}^{(l)}\beta_{.1}^0 - D_{13}^{(l)}\gamma_{.1}), \tag{C.16}$$

where

$$W_{11}^{(l)} = M_{11}^{(l)} - [M_{13}^{(l)} \quad M_{14}^{(l)}]\begin{bmatrix} M_{33}^{(l)} & M_{34}^{(l)} \\ M_{43}^{(l)} & M_{44}^{(l)} \end{bmatrix}^{-1}\begin{bmatrix} M_{31}^{(l)} \\ M_{41}^{(l)} \end{bmatrix}. \tag{C.17}$$

$\partial W_{11}^{(l)}/\partial l$ can be computed by numerical methods

$$S_{\beta^0_{.1}l} = -\sigma_{11} \operatorname*{plim}_{T\to\infty} \frac{1}{T} \frac{\partial^2 L}{\partial \beta^0_{.1}\, \partial l} = \left(\dot{S}_{\beta_{.1}l}\right), \qquad (C.18)$$

$$\frac{\partial L^*}{\partial \sigma_{11}} = -\frac{T}{2} \cdot \frac{1}{\sigma_{11}} + \frac{T}{2} \cdot \frac{1}{\sigma_{11}^2}\, (\beta^{0'}_{.1} M_{11}^{(l)} \beta^0_{.1} - 2\beta^{0'}_{.1} M_{13}^{(l)} \gamma_{.1} + \gamma'_{.1} M_{33}^{(l)} \gamma_{.1}),$$
$$(C.19)$$

$$\frac{\partial^2 L^*}{\partial \sigma_{11}^2} = \frac{T}{2\sigma_{11}^2} - \frac{T}{\sigma_{11}^3}\, (\beta^{0'}_{.1} M_{11}^{(l)} \beta^0_{.1} - 2\beta^{0'}_{.1} M_{13}^{(\lambda)} \gamma_{.1} + \gamma'_{.1} M_{33}^{(l)} \gamma_{.1}),$$
$$(C.20)$$

$$S_{\sigma_{11}\sigma_{11}} = -\frac{1}{2\sigma_{11}} + \frac{1}{\sigma_{11}} = \frac{1}{2\sigma_{11}}, \qquad (C.21)$$

$$\frac{\partial^2 L^*}{\partial \sigma_{11}\, \partial l} = \frac{T}{2\sigma_{11}^2}\, (\beta^{0'}_{.1} D_{11}^{(l)} \beta^0_{.1} - 2\beta^{0'}_{.1} D_{13}^{(l)} \gamma_{.1} + \gamma'_{.1} D_{33}^{(l)} \gamma_{.1}), \qquad (C.22)$$

$$S_{\sigma_{11}l} = -\frac{T}{2\sigma_{11}} \operatorname*{plim}_{T\to\infty} (\beta^{0'}_{.1} D_{11}^{(l)} \beta^0_{.1} - 2\beta^{0'}_{.1} D_{13}^{(l)} \gamma_{.1}$$
$$+ \gamma'_{.1} D_{33}^{(l)} \gamma_{.1}), \qquad (C.23)$$

$$\frac{\partial L^*}{\partial l} = -\frac{T}{2} \left\langle 2(W^{(l)})^{-1} - \operatorname{diag}(W^{(l)})^{-1}, \frac{\partial W^{(l)}}{\partial l} \right\rangle$$
$$+ \frac{T}{2} \cdot \frac{\beta^{0'}_{.1}(\partial W_{11}^{(l)}/\partial l)\beta^0_{.1}}{\beta^{0'}_{.1} W_{11}^{(l)} \beta^0_{.1}}$$
$$- \frac{T}{2} \frac{1}{\sigma_{11}} (\beta^{0'}_{.1} D_{11}^{(l)} \beta^0_{.1} - 2\gamma'_{.1} D_{31}^{(l)} \beta^0_{.1} + \gamma'_{.1} D_{33}^{(l)} \gamma_{.1}), \quad (C.24)$$

where (A, B) is the inner product in the $[G(G + 1)]/2$-dimensional space of $G \times G$ symmetric matrices.

The term $\Sigma_{t=1}^{T} \Sigma_{i=1}^{M} \ln y_{ti}$ in (81) vanishes due to normalization of the variables:

$$\frac{\partial^2 L^*}{\partial l^2} = -\frac{T}{2} \frac{\partial^2}{\partial l^2} \ln |W^{(l)}|$$
$$+ \frac{T}{2} \cdot \frac{\beta^{0'}_{.1} \frac{\partial^2 W_{11}^{(l)}}{\partial l^2} \beta^0_{.1}}{\beta^{0'}_{.1} W_{11}^{(l)} \beta^0_{.1}} - \frac{T}{2} \cdot \frac{\left(\beta^{0'}_{.1} \frac{\partial W_{11}^{(l)}}{\partial l} \beta^0_{.1}\right)^2}{(\beta^{0'}_{.1} W_{11}^{(l)} \beta^0_{.1})^2}$$
$$- \frac{T}{2} \cdot \frac{1}{\sigma_{11}} \left(\beta^{0'}_{.1} \frac{\partial D_{11}^{(l)}}{\partial l} \beta^0_{.1} - 2\gamma'_{.1} \frac{\partial D_{31}^{(l)}}{\partial l} \beta^0_{.1} + \gamma'_{.1} \frac{\partial D_{33}^{(l)}}{\partial l} \gamma_{.1}\right).$$
$$(C.25)$$

The first and second derivatives of $W^{(l)}$ and $W_{11}^{(l)}$ will be calculated by numerical methods. We also have

$$S_{ll} = -\sigma_{11} \plim_{T \to \infty} \left(\frac{1}{T} \frac{\partial^2 L^*}{\partial l^2} \right). \tag{C.26}$$

ACKNOWLEDGMENTS

I wish to acknowledge the contributions of Chi Wing Wong, Steve Peters, Richard Petersen, and Allan Cheadle, who worked with me at various stages of the research on this project. I also wish to acknowledge the contribution of the staff of the Sacramento Municipal Utility District, particularly Charles Hair, Freeman Dom, and Jack Smith. Lester Taylor and Jerry Hausman read the entire manuscript. Phoebus Dhrymes read Sections VI.A–D and Appendices A and C, and Dennis Aigner read various parts of the manuscript. All made valuable comments. Many of their suggestions for revision have been incorporated into this work. Needless to say, any remaining errors are mine. I also wish to thank Edythe J. Khazzoom, Dolores Armada, and Michelle Moretti for typing and Patricia Murphy and Josef Chytry for inputting the manuscript. The support provided by the Electric Power Institute for this project is gratefully acknowledged.

REFERENCES

Allen, R. G. D. (1975). "Index Numbers in Theory and Practice." Aldine, New York.

Amemiya, T. (1980). Selection of regressors. *Int. Economic Rev.,* June, 331–354.

Amemiya, T. (1974). The nonlinear two-stage least-squares estimator. *J. Econometrics* **2,** 105–110.

Amemiya, T., and Powell, J. L. (1981). A comparison of the Box–Cox maximum likelihood estimator and the nonlinear two-stage least-squares estimator. *J. Econometrics* **17,** 351–381.

Applebaum, E. (1979). On the choice of functional forms. *Int. Economic Rev.,* June, 449–458.

Basmann, R. L. (1960). On finite sample distributions of generalized classical linear identifiability test statistics. *J. Am. Stat. Assoc.,* December, 650–659.

Berndt, E. (1978). The Demand for Electricity: A Comment and Further Results. Resources Paper No. 28, Department of Economics, University of British Columbia, Xerolith.

Bickel, P. J., and Doksum, K. A. An analysis of transformations revisited. *J. Am. Stat. Assoc.* June, 296–311.

Bickel, P. J., and Doksum, K. A. (1983). "An Analysis of Transformations Revisited II." Department of Statistics, University of California at Berkeley, Berkeley, California.

Box, G. E. P., and Cox, D. R. (1964). An analysis of transformations. *J. R. Stat. Soc. Ser. B,* **26,** 211–243.

Box, G. E. P., and Cox, D. R. (1982). An analysis of transformations revisited, rebutted. *J. Am. Stat. Assoc.* March, 209.

Breiman, L., and Freedman, D. (1983). How many variables should be entered in a regression equation? *J. Am. Stat. Assoc.,* March, 131–136.

Breusch, T. S., and Pagan, A. R. (1980). The Lagrange multiplier test and its applications to model specification in econometrics. *Rev. Econom. Stud.* **XLVII,** 239–253. .

Burtless, G., and Hausman, J. J. (1978). The effect of taxation on labor supply: Evaluation of the negative income tax experiment. *J. Political Econ.,* December, 1103–1130.

Buse, A. (1982). The likelihood ratio, Wald, and Lagrange multiplier tests: an expository note. *Am. Stat.,* August 153–157.

Carroll, R. J. (1985). Tests for regression parameters in power transformation models. *Scand. J. Stat.* (in press).

Carroll, R. J., and Ruppert, D. (1981). On prediction and the power transformation family. *Biometrika* **68,** 609–617.

Carroll, R. J., and Ruppert, D. (1982). "Power Transformations when Fitting Theoretical Models to Data." Technical Report, University of North Carolina, Chapel Hill, North Carolina.

Chang, H.-S. (1977). Functional forms and the demand for meat in the United States. *Rev. Economics Stat.,* August, 355–359.

Clark, J. M. (1955). Competition: Static models and dynamic aspects. *Am. Economic Rev.,* May, 450–462.

Dhrymes, P. J. (1970). "Econometrics: Statistical Foundations and Applications." Harper, New York.

Dhrymes, P. J. "Introduction to Econometrics." Springer-Verlag, Berlin and New York.

Dhrymes, P. J., and Kurz, M. (1964). Technology and scale in electricity generation. *Econometrica,* July, 287–315.

Doksum, K. A., and Wong, C. W. (1983). Statistical tests based on transformed data. *J. Am. Stat. Assoc.,* June, 411–417.

Draper, N. R., and Cox, D. R. (1969). On distributions and their transformation to normality. *J. R. Stat. Soc. Ser. B* **31,** 291–324.

Dubin, J. A. (1981). "Rate Structure and Price Specification in the Demand for Electricity." Department of Economics, Massachusetts Institute of Technology.

Dubin, J. A., and McFadden, D. (1980). "An Econometric Analysis of Residential Appliance Holdings and Consumption." Department of Economics, Massachusetts Institute of Technology.

Durbin, J. (1957). Testing for serial correlation in systems of simultaneous regression equations. *Biometrika,* December, 370–377.

Durbin, J., and Watson, G. S. (1951). Testing for serial correlation in least-square regression, II. *Biometrika,* June, 159–178.

Efron, B. (1982). "The Jacknife, the Bootstrap and Other Resampling Plans. Society of Industrial and Applied Mathematics, Philadelphia, Pennsylvania.

Evans, G. B. A., and Savin, N. E. (1982). Conflict among the criteria revisited: The W, LR and LM test. *Econometrica,* May, 737–748.

Freedman, D. A. (1981). Bootstrapping regression models. *Ann. Stat.* **9,** 1218–1228.

Freedman, D. A., and Peters, S. C. (1982). "Bootstrapping A Regression Equation: Some Empirical Results." Revised, Technical Report # 10, Department of Statistics, University of California at Berkeley, Berkeley, California.

Freedman, D. A., and Peters, S. C. (1983). "Bootstrapping an Econometric Model: Some Empirical Results." Technical Report No. 21. Department of Statistics, University of California at Berkeley, Berkeley, California.

Galbraith, J. K. (1967). "The New Industrial State." Cambridge University Press, New York and London.

Godfrey, L. G. (1978). A note on the use of Durbin's *h* test when the equation's estimated by instrumental variables. *Econometrica,* January, 225–228.

Goldberg, S. (1961). "Difference Equations." Wiley, New York.

Goldburger, A. S. (1972). Maximum-likelihood estimation of regressions containing unobservable independent variables. *Int. Economic Rev.* February, 1–15.

Halvorsen, R. (1978). "Econometric Models of U.S. Energy Demand." D.C. Heath and Co., Lexington, Massachusetts.

Hausman, J. A. (1979). Individual discount rates and the purchase and utilization of energy using durables. *Bell J. Economics,* Spring, 33–54.

Hausman, J. A., Kinnucan, M., and McFadden, D. (1979). A two-level electricity demand model; evaluation of the Connecticut time-of-day pricing test. *J. Econometrics* **10,** 263–289.

Hinkley, D. V. (1975). On power transformation to symmetry. *Biometrika* **62**(1), 101–111.

Judge, G. G., Griffiths, W. E., Hill, R. C., and Lee, T. C. (1980). "The Theory and Practice of Econometrics." Wiley, New York.

Khazzoom, J. D. (1971). The FPC staff econometric model of the U.S. natural gas supply. *Bell Econ. Manage. Sci.,* Spring, 51–93.

Khazzoom, J. D. (1984). "The Demand for Home Insulation: A Study in the Household Demand for Conservation." Working Paper no. EAP-8, Berkeley Business School, University of California, Berkeley, California.

Khazzoom, J. D. (1986). "Integrating Conservation Measures in the Estimation of the Demand for Electricity." JAI Press, forthcoming.

Kmenta, J. (1971). "Elements of Econometrics." Macmillan, New York.

Leech, D. (1975). Testing the error specification in nonlinear regression. *Econometrica,* July, 719–725.

McFadden, D., Puig, C., and Kirshner, D. (1977). *Determinants of the Long Run Demand for Electricity.* Mimeo.

Mukerji, V. (1963). A Generalized SMAC function with constant ratios of elasticities of substitution. *Rev. Econ. Studies,* pp. 233–236.

Nordin, J. A. (1976). A proposed modification of Taylor's demand analysis: comment. *Bell J. Econ.* Autumn, 719–721.

Pacific Gas and Electric Company (1982a). "RCS Follow-up Survey: MR-82-2." PG&E, San Francisco, California.

Pacific Gas and Electric Company (1982b). "ZIP Follow-Up Survey: MR-82-1." PG&E, San Francisco, California.

Plosser, C. I., Schwert, G. W., and White, H. (1982). Differencing as a test of specification. *Int. Econ. Rev.,* October 535–552.

Poirier, D. J. (1978). The use of the Box–Cox transformation in limited dependent variable models. *J. Am. Stat. Assoc.,* June, 284–287.

Ramsey, J. B., and Schmidt, P. (1976). Some further results on the use of OLS and BLUS residuals in specification error tests. *J. Am. Stat. Assoc.,* June, 389–390.

Rosen, H. S. (1976). Taxes in a labor supply model with joint wage-hours determination. *Econometrica,* May, 485–507.

Sacramento Municipal Utility District (1978). "Common Forecasting Methodology II." Technical Documentation, SMUD, Sacramento, California.

Sacramento Municipal Utility District (1980). "1979 Annual Report: Energy Conservation Activities." SMUD, Sacramento, California.

Sacramento Municipal Utility District (1981). "1980 Annual Report: Energy Conservation Programs and Activities." SMUD, Sacramento, California.

Sacramento Municipal Utility District (1982). "1981 Annual Report: Energy Conservation Activities." SMUD, Sacramento, California.

Sargan, J. D. (1980). Some tests of dynamic specification for a single equation. *Econometrica,* May, 879–897.

Savin, N. E., and White, K. J. (1978). Estimation and testing for functional form and autocorrelation: A simultaneous approach. *J. Econometrics,* August, 1–12.

Savin, N. E., and White, K. J. (1977). The Durbin–Watson test for serial correlation with extreme sample sizes or many regressors. *Econometrica,* November, 1989–1996.

Schlesselman, J. (1971). Power families: A note on the Box and Cox transformation." *J. R. Stat. Soc. Ser. B* **33,** 472–476.

Schumpeter, J. A., (1934). "The Theory of Economic Development." Harvard University Press, Cambridge, Massachusetts.

Shapiro, S. S., and Wilk, M. B. (1965). An analysis of variance test for normality (complete samples). *Biometrika,* December, 591–611.

Southern California Edison (1982). "1981 Conservation and Load Management, Vol. II, Residential Program Evaluation." Rosemead, California.

Spitzer, J. (1976). The demand for money, the liquidity trap, and functional forms. *Int. Economic Rev.,* February, 220–227.

Spitzer, J. (1977). A simultaneous equations system of money demand and supply using generalized functional forms. *J. Econometrics,* January, 117–128.

Spitzer, J. (1978). A Monte Carlo investigation of the Box–Cox transformation in small samples. *J. Am. Stat. Assoc.* September, 488–495.

Spitzer, J. (1982). Primer on Box–Cox estimation. *Rev. Economics Stat.,* May, 307–313.

Stuval, G. (1957). A new index number formula. *Econometrica,* January, 123–131.

Taylor, L. D. (1975). The demand for electricity: A survey. *Bell J. Economics,* Spring, 74–110.

Taylor, L. D., Blattenberger, G. R., and Verleger, P. K. (1977). "The Residential Demand for Energy. Electric Power Research Institute, EA-235, Palo Alto, California.

Taylor, L. D., Blattenberger, G. R., and Pennhack, R. K. (1982). "The Residential Demand for Energy." Electric Power Research Institute, EA-1572, Palo Alto, California.

Wales, T. J., and Woodland, A. D. (1979). Labor supply and progressive taxes. *Rev. Econ. Stud.* January, 83–95.

Weiss, C. S., and Newcomb, T. M. (1982). Evaluation of the home energy check program. *In* EPRI, Proceedings of a Workshop at Columbus, February, Draft.

Wong, C. W. (1984). "Transformation of Independent Variables in Regression Models." Ph.D. dissertation, University of California, Berkeley, California.

Zarembka, P. (1974). Transformation of variables in econometrics. *In* "Frontiers in Econometrics " (P. Zarembka, ed.). Academic Press, New York.

Subject Index

313